Palgrave Studies of Marketing in Emerging Economies

Series Editors

Robert E. Hinson
Ghana Communication Technology University
Accra, Ghana

Ogechi Adeola
Pan-Atlantic University
Lagos Business School
Lagos, Nigeria

This book series focuses on contemporary themes in marketing and marketing management research in emerging markets and developing economies. Books in the series cover the BRICS (Brazil, Russia, India, China and South Africa), MINT (Mexico, Indonesia, Nigeria and Turkey), CIVETS (Colombia, Indonesia, Vietnam, Egypt, Turkey, and South Africa); EAGLE economies (those which are expected to lead growth in the next ten years, such as Brazil, China, India, Indonesia, South Korea, Mexico, Russia, Taiwan, and Turkey) and all other African countries (classified under developing countries), taking into consideration the demographic, socio-cultural and macro-economic factors influencing consumer choices in these markets. The series synthesizes key subject areas in marketing, discuss marketing issues, processes, procedures and strategies across communities, regions and continents, and also the way digital innovation is changing the business landscape in emerging economies.

Palgrave Studies of Marketing in Emerging Economies presents a unique opportunity to examine and discuss marketing strategy and its implications in emerging economies, thereby filling a gap in current marketing literature.

All chapter submissions to the series will undergo a double blind peer review and all book proposals will undergo a single blind peer review.

Asphat Muposhi • Brighton Nyagadza
Editors

Sustainable Green Marketing Strategies for a Circular Economy

African Realities and Imperatives

palgrave
macmillan

Editors
Asphat Muposhi
Marketing Division
School of Business Sciences
University of the Witwatersrand
Johannesburg, South Africa

Brighton Nyagadza
School of Business and Health Studies
York St. John University (YSJU)
London, UK

ISSN 2730-5554 ISSN 2730-5562 (electronic)
Palgrave Studies of Marketing in Emerging Economies
ISBN 978-3-031-81626-0 ISBN 978-3-031-81627-7 (eBook)
https://doi.org/10.1007/978-3-031-81627-7

This Palgrave Macmillan imprint is published by the registered company Springer Nature Switzerland AG.
The registered company address is: Gewerbestrasse 11, 6330 Cham, Switzerland

Acknowledgement

Our foremost gratitude goes to Palgrave Macmillan for initiating an academic discourse on the status of green marketing in Africa. We believe that this African perspective will significantly contribute to the evolving discipline of green marketing. We are also grateful for the support we received from the Palgrave Macmillan Editorial team which was critical for the success of this project. We also extend our profuse thanks to various scholars who contributed to this book. Your rich insights improved the quality of this book. Finally, we are also indebted to the anonymous reviewers who provided constructive feedback that enhanced the quality of this book.

About this Book

As the global economy evolves, managing the sanctity of the natural environment and promoting sustainable economic growth are the major challenges confronting humanity. Increasingly, green marketing is being positioned as a strategic imperative for fostering sustainable development. Despite this, mixed views are prevalent in developing economies on the role of green marketing as a pathway to deliver transformative and structural changes required to stimulate sustained economic growth. This book contributes to the evolving discourse on green marketing by exploring the realities and imperatives necessary to implement sustainable green marketing strategies in developing economies. This book adopts a multidisciplinary approach that integrates the micro and macro perspectives to provide insights into the realities and imperatives that facilitate or inhibit the realisation of the sustainable green marketing agenda in Africa. The contestations associated with the governance and practice of green marketing are discussed. The role of environmental values, norms and beliefs are interrogated.

This book also examines the willingness of consumers to support sustainable consumption practices, such as collaborative consumption, sharing economy, consumption of green products and intention to engage in circular economy practices such as sustainable tourism, green reverse logistics and eco-innovation. The promotion of green marketing strategies is situated within the fourth industrial revolution trajectory. Thus, the role of artificial intelligence and social media in fostering sustainable production and consumption is also discussed. There are mixed views on the sustainability of green marketing as a business model in emerging economies. This book intends to contribute to this debate by examining the value proposition of green marketing as a business model. The practice of green marketing and the progress thereof can only be measured through transparent reporting of sustainability indicators. In this regard, incidences of under-reporting and greenwashing have been reported. This book also examines the role of sustainability indicators in promoting accountability in the practice of green marketing. This book offers valuable insights to researchers and policymakers who intend to develop and implement sustainable green marketing strategies in developing economies. Marketing managers are also empowered with input to formulate effective value propositions that promote the mainstream adoption of green marketing practices. This book also enriches consumers' understanding of green marketing with a focus on its environmental, economic and social contributions.

Contents

Notes on Contributors

Emmanuel Chao, PhD is Associate Professor affiliated with the School of Business at Mzumbe University. He is currently the manager for Innovation, Entrepreneurship and Industrial Linkage at Mzumbe University, Director of the Tanzania Marketing Science Association, chair of the ICBMEM conference, and editor of the *Journal of Applied Marketing Science*.

Munyaradzi Chibaro (PhD) is a Lecturer in the Department of Management at University of Botswana. His research interests are in supply chain management and green reverse logistics. he has published several articles in peer reviewed journals.

Tendai Chibaya holds a PhD in Economics and Management Sciences with Tourism Management from North West University, South Africa. She lectures in the Department of Hospitality, Tourism and Culture at Great Zimbabwe University, Zimbabwe. Her research interests are in event tourism management, cultural and heritage tourism, disability tourism and tourism management.

Maximillian Chikwanha is Senior Lecturer in Marketing and Module Leader: Marketing Theory and Practice at Leicester Castle Business School, De Montfort University, Leicester. She is also Deputy Programme leader: UG Business and Marketing as well as BAL BAME Lead: Lead and Inspire Project.

David Chikwere, PhD is an accredited Chartered Sustainable Supply Chain and Logistics Professional (CSSCLP) as well as a Chartered

Procurement and Supply Professional (CIPS). His experience spans over 15 years of training and supply chain business with demonstrated experience in procurement planning, logistics, supplier relationship management, and lecturing. His research interest covers public procurement, circular economy, global supply chain transformation, and sustainability.

Sharon Chisango is a full-time lecturer in the Human Capital Management Department at the Bindura University of Science Education. She is a holder of a DPhil in Human Resources Management, an MSc in Human Resources Management, and a BBS in Human Resources Management. She also holds a Postgraduate Diploma in Education. Her research interests are green human resources management and public sector administration.

Thulani Dube is a full-time lecturer in the Department of Marketing at the Bindura University of Science Education. He holds a PhD in Management, an MBA as well as a BBA. His research interests are in digital marketing.

Mufaro Dzingirai (Ph.D) is a Lecturer in the Department of Business Management at Midlands State University. His research interests include Digital Transformation, Supply Chain Management, Strategy, Management, Finance, Entrepreneurship, and Development.

Shelton Giwa, PhD is Subject Lead & Senior Lecturer at Leicester Castle Business School, De Montfort University, Leicester. Shelton's research lies at the intersection of pedagogy (student experience as well as assessment and feedback), digital marketing (e.g. mobile marketing), and consumer response.

Hazel Gwasira is an avid independent Evironmentalist and Philanthropist. She has worked with several non-governmental organisations promoting environmental sustainability and committed to the upliftment of disadvantaged young girls in Zimbabwe.

Divaries Cosmas Jaravaza is a full-time lecturer in the Department of Marketing at the Bindura University of Science Education. He is a holder of a PhD, an MBA, M Com in Marketing Strategy, and a BBA in Marketing. His research interests include, but are not limited to, social marketing, health care marketing, and green marketing.

Tatenda Katsande is a full-time lecturer in the Business Management Department at the Lupane State University. She holds an MSc in Marketing, an MSc in Supply Chain Management, and a BSc in International Marketing. Her research interests are women's healthcare marketing and green marketing.

Fungai Jacqueline Kiwa, PhD has expertise in AI, IoT framework design, and programming in Java, Python, and C++. With a passion for harnessing technology to drive positive change, she has authored numerous publications and presentations, showcasing her expertise in creative problem-solving and cutting-edge solutions. As a seasoned professional, she is a respected member of prestigious technology associations, including IEEE, SWE, IET, and ACM.

Siphiwe Plaxcedes Mandina, PhD is a lecturer in Marketing at Leicester Castle Business School, De Montfort University. She specialises in Strategic Brand Management, Contemporary Marketing Practice, and Tourism Marketing. Her academic pursuits are complemented by practical industry marketing experience, making her a valued voice in both educational and professional settings.

Herbert Masukume is a Lecturer in the Faculty of Business Sciences at Midlands State University. He holds an MSc in Strategic Management from Great Zimbabwe University and a BSc in Human Resources Management from Midlands State University (Zimbabwe). His interests are in Industrial Relations with a focus on improving workplace relations for the maintenance of industrial harmony.

Dumisani Mawonde is a Lecturer in the Department of Logistics, Faculty of Business and Economic at Nelson Mandela University. His research interests are in green supply chain management. He has published several articles in peer reviewed journals.

Martin Muduva holds an MSc degree in Entrepreneurship and Innovation, an MSc in Mechatronics and AI, a Master of Leadership and Corporate Governance, and an MSc in Big Data Analytics, as well as a B. Tech (Hons) in Information Security and Assurance. His multidisciplinary expertise makes him a valuable asset in technology, innovation, and corporate governance.

Chipo Mukonza, DTech is a business development consultant, market research expert, and clean cooking advocate. Her research focuses on the intersection of climate change, marketing, adaptive capacity, and clean energy. She passionately believes that companies and communities need to understand and believe in the benefits of sustainability for their businesses.

Paul Mukucha is a lecturer in the Department of Marketing at the Bindura University of Science Education. He holds a PhD in Business Management, MBA, MSc in Supply Chain Management, B. Com (hons) Supply Chain Management, B. Com (hons) Marketing, and a PGDTE. His research interests are in Supply Chain Management, Green Marketing, and Social Marketing.

Shakerod Munuhwa, PhD is Lecturer in Operations and Supply Chain Management at Glasgow Caledonian University in UK. He has a PhD in Logistics from Nelson Mandela University in South Africa. He is a chartered member of the CILT with extensive previous experience in Contracts logistics and beverages distribution. He has a strong research focus on sustainable supply chain management and supply chain digitalisation.

Asphat Muposhi, PhD, DTech is a post doc research fellow in the Marketing Division, School of Business Sciences, University of the Witwatersrand, Johannesburg, South Africa. He has published several articles in international peer-reviewed journals published by Sage, Cambridge, and Inderscience. His research interests are green consumerism, environmental sustainability, and environmental education.

Chipo Mutongi, PhD is a lecturer at the Midlands State University (MSU) with more than sixteen years of lecturing experience. She is Associate Editor in Chief for the *International Journal of Doctoral Studies* and a member of IAS, IODZ, IPMZ, and ZIMLA. She has published several articles in peer-reviewed international journals and SCOPUS-indexed journals and presented at various international conferences.

Brighton Nyagadza is a CABS (UK) Certified Business and Management Educator (CMBE) who holds a Ph.D. in Marketing. He is a full-time Digital Marketing academic and Unit of Assessment Lead for Business and Management Studies at York St. John University (YSJU), London Campus, England, United Kingdom; a full member of the Marketers Association of Zimbabwe (MAZ); an Associate of The Chartered Institute of Marketing (CIM), United Kingdom; and Power Member of the Digital

Marketing Institute (DMI), Dublin, Ireland. He has published books with Routledge of Taylor & Francis Group, CAB International United Kingdom (UK), Palgrave Macmillan, and reputable international journals such as *Journal of Digital Media & Policy* (Intellect), *Sustainable Technology & Entrepreneurship* (Elsevier), *European Journal of Innovation Management* (Emerald), *Journal of Entrepreneurship In Emerging Economies* (Emerald), *Youth & Society* (SAGE), *PSU Research Review* (Emerald), *Journal of Fashion Marketing & Management* (Emerald), *Journal of Environmental Media* (Intellect), *Africa Review* (Brill), *Tourism Critiques: Practice & Theory* (Emerald), *Quality Assurance in Education* (Emerald), *International Journal of Public Sector Management* (Emerald), *The Marketing Review* Westburn), *Journal of Asian & African Studies* (SAGE), among others. Brighton serves on various international boards including the editorial board of *Journal of Digital Media & Policy* and *The Retail and Marketing Review*.

Theo Tsokota, PhD is a distinguished academic at the Midlands State University in Zimbabwe. His contributions continue to resonate within the fields of technology, education, and tourism, making him a key figure in the discourse on sustainable development in Zimbabwe and beyond. His research continues to shape policies and practices in Zimbabwe and internationally.

Zibanai Zhou is a lecturer in the Department of Tourism, Hospitality and Leisure Sciences at Midlands State University, Zimbabwe. He has over 15 years teaching experience at institutions of higher learning. His main research interests span: Tourism development, event tourism, and local communities. Zibanai has published several book chapters with Springer, Routledge Francis and Taylor, Palgrave MacMillan, and CABI. He has also published in international peer reviewed journals.

Melissa Zulu, PhD is the head of the Marketing Department in the School of Business Sciences at the University of the Witwatersrand, Johannesburg, South Africa. Her research interests are Data Science in Marketing, Retail Management, Services Marketing, Entrepreneurial Marketing, and Township Economy.

Kudakwashe Zvitambo (Ph.D) is a Senior Lecturer and Deputy Dean in the Faculty of Business Sciences at Midlands State University in Zimbabwe. He is a prolific researcher who publishes in highly reputable journals. His research interests are corporate governance, strategic leadership and management, sustainable development, digital transformation, and green economy.

List of Figures

List of Tables

CHAPTER 1

Sustainable Green Marketing Strategies for a Circular Economy in Africa: An Introduction

Asphat Muposhi and Brighton Nyagadza

Introduction

Green marketing is gaining prominence as a strategy that directs national governments' response to environmental sustainability challenges (Mukonza & Swarts, 2020). It involves the production, consumption, distribution and disposal of products and services in a manner that is beneficial to the natural environment (Sharma & Kushwaha, 2019). Green

A. Muposhi (✉)
Marketing Division, School of Business Sciences, University of the Witwatersrand, Johannesburg, South Africa
e-mail: asphat.muposhi@wits.ac.za

B. Nyagadza
School of Business and Health Studies, York St. John University (YSJU), London, UK
e-mail: b.nyagadza@yorksj.ac.uk

A. Muposhi, B. Nyagadza (eds.), *Sustainable Green Marketing Strategies for a Circular Economy*, Palgrave Studies of Marketing in Emerging Economies,
https://doi.org/10.1007/978-3-031-81627-7_1

marketing is operationalised under the concept of sustainability with a focus on economic, social and environmental pillars (Mondal et al., 2022). The central principle of the sustainability concept is to strive to meet immediate human needs in a manner that does not compromise the ability of future generations to sustain their own livelihoods (Mendoza et al., 2021). In developing economies, green economic growth and improved quality of life are used as the main value propositions of the green marketing agenda (Jakob et al., 2020). As the threats of climate change are becoming more noticeable coupled with a shrinking global economy, there is a growing perception that the concept of green marketing has failed to deliver on its key objectives (Sharma, 2021). It is against this background that this book interrogates the green marketing promise in the context of developing economies in Africa.

Green Marketing Governance

As the world grapples with the challenges of environmental degradation and climate change, indigenous wisdom offers a wealth of knowledge that can inform sustainable green marketing practices (Terblanché-Greeff, 2019). The implementation of green marketing practices has been mainly driven by environmental regulations and commitments to environmental treaties (Asongu & Odhiambo, 2021; Sabela-Rikhotso et al., 2021). The major weakness of this approach has been the operational and economic burden associated with enforcement of environmental laws (Sabela-Rikhotso et al., 2021). Moreover, the commitment of national governments to green marketing issues often depends on the political ideology of the governing party (Cruz, 2017). In the majority of African countries, a transition to a green economy is regarded as not urgent and is perceived as an antithesis to economic growth. As a result, self-regulated behaviour governed by social norms, values and beliefs has emerged as another form of green marketing governance (Jakob et al., 2020). Against this backdrop, this book examines the role of indigenous knowledge systems in green marketing governance, with a focus on the role of environmental worldviews, values, beliefs and norms. Drawing from the African philosophy of Ubuntu, the role of indigenous knowledge systems as tools to promote green marketing practice in a way that enhances environmental justice and improves quality of life is interrogated.

Green Marketing as a Business Model

In corporate literature, green marketing is positioned as a sustainable business model (Mukonza & Swarts, 2020; Swinkels, 2024). Green marketing represents a paradigm shift in business approaches towards achieving sustainable development and a greener economy (Mukonza & Swarts, 2020). This paradigm, which gained prominence in the late twentieth century, emerged as a response to the need to strike a balance between the competing interests of economic growth and environmental sustainability. As a business model, green marketing is expected to stimulate economic growth through eco-innovations and eco-preneurship (Mendoza et al., 2021). The acceptance of social eco-innovations is also examined in this book.

Although green marketing has the potential to spur economic growth and improve the quality of life, there is a growing concern in developing economies that green marketing is evolving into green capitalism (Bosch & Schmidt, 2019; Mendoza et al., 2021). In developing economies, especially of African strand, the implementation of green marketing is characterised by lack of environmental standards as well as lapses in monitoring and enforcing environmental laws (Sabela-Rikhotso et al., 2021). Exploiting this reality, big companies are implementing self-regulated, voluntary green marketing initiatives such as green products, sustainable packaging and carbon credits which are often criticised for being skewed towards profitability (Mendoza et al., 2021; Swinkels, 2024). Green marketing is also blamed for recalibrating the notions of capitalism in the form of the exorbitant price of green products which is not aligned with the pro-poor African policies (Mendoza et al., 2021). This book contributes to the foregoing insights by assessing the sustainability of green marketing as a business model. Strategies that can be employed by policymakers to reconcile green marketing and perceptions of green capitalism are also explored. The role of green corporate governance as a strategy for managing the corporatisation of green marketing is discussed.

Green Marketing Liability

As the concept of green marketing evolves, critics are cautioning on what has been dubbed the 'green marketing rhetoric' due to the continued depressed levels of economic growth, higher levels of environmental degradation, poverty and unemployment (Sabela-Rikhotso et al., 2021; Aluko et al., 2021). Energy poverty still persists as attested by ongoing load shedding in most African economies while the promise of green jobs is fading due to lack of resources to fund eco-innovations that underpin a transition to a circular economy (Jha & Schmidt, 2021). In most African economies, the energy mix remains skewed towards fossil fuels with investment in renewable energy such as solar, wind and biofuels lagging behind due to policy inertia (Jha & Schmidt, 2021). African economies are still rated high on carbon emissions. It is estimated that African economies contribute 3 percent to the world's total carbon dioxide emissions and remain exposed to climate change-induced challenges such as droughts and floods (Sakariyahu et al., 2023). Consequently, an understanding of metrics that can be used to evaluate the success of green marketing strategies is a key focus area for this book. Green marketing reporting is regarded as fragmented in Africa (Jeannotte & Dick, 2024). This book therefore provides input towards the development of an integrated green marketing reporting system.

Circular Economy Transition Through Green Supply Chain Management

The transition to a circular economy largely depends on the extent to which value chain players are implementing green marketing practices (Mishra et al., 2017; Agyabeng-Mensah et al., 2020). Green supply chain management has emerged as a vital strategy for promoting the implementation of green marketing practices in the value chain (Namagembe et al., 2019). Green supply chain management incorporates environmental considerations in key stages of the value chain such as product design, procurement of raw materials, production procedures, final product delivery and end-of-life management (Seman et al., 2019; Wang et al., 2020). Despite the critical role of green supply chain management in fostering green marketing, there is a lack of studies that comprehensively address the interaction between these two elements in an African context. This book contributes to address this gap by examining how green supply chain management is advancing the green marketing agenda in Africa. The role of green reverse logistics practices such as recycling, remanufacturing and refurbishment among others as strategies for facilitating a transition to a circular economy is also examined.

Green Marketing and Digital Economy

The prevalence of greenwashing is blamed for fostering ambivalent consumer attitudes towards benign green products (Gatti et al., 2021). In order to promote mainstream consumption of green products, Mkhize and Ellis (2024) underscore the importance of developing effective green marketing strategies. A key focus area in developing green marketing strategies involves the greening of the marketing mix in a manner that promotes sustainable consumption and a transition to a circular economy. To do this, marketers of green products may need to tap into the opportunities presented by the growth in digital marketing. The digital approach gains support from emerging literature that posits that digital platforms can be leveraged to address environmental problems such as loss of biodiversity, waste and pollution (Schöggl et al., 2023). This approach gains support from the United Nations Environmental Programme's digital platform, and the World Environment Situation Room. The African continent presents peculiar challenges due to the prevalence of the digital divide and variation in consumers' technology literacy levels. This book contributes to the evolving debate on the nexus between green marketing and digital economy by examining the role of artificial intelligence-driven green marketing strategies in promoting a circular economy. The effectiveness of social media nudging as a strategy for disseminating sustainable end-of-life product disposal information is also examined.

Book Structure

This book has 13 chapters which are organised as follows: the introductory chapter highlights the evolution of the concept of green marketing with a focus on governance, its potential as a business model, green supply chain management, evaluation of green marketing strategies and the role of the digital economy. Chapter 2 discusses the role of African indigenous knowledge management systems in fostering green marketing governance with a focus on values, norms and beliefs. This chapter argues that by unlocking the power of indigenous wisdom, policymakers can develop more effective and sustainable green marketing strategies that promote environmental justice and improve quality of life. Chapter 3 provides a systematic literature review on green corporate governance and sustainable development goals. It proposes lessons for Africa from best practices gleaned from the Global North. Chapter 4 examines the role of sustainability indicators as tools for evaluating green marketing strategies. Chapter 5 examines the role of eco-innovations as a circular economy practice with a focus on reusable sanitary towels.

With the aid of a multi-case study analysis, Chap. 6 examines the efficacy of green marketing as a sustainable business model. Using quantitative data, Chap. 7 examines consumer perspectives related to the marketing of green products. Acknowledging that the tourism and hospitality industry is one of the major sectors with higher levels of carbon footprint, Chap. 8 discusses the role of sustainable tourism as a green marketing practice with a focus on tourism heritage sites. Chapter 9 reviews literature on the intersection between green marketing and green supply chain management. Chap. 10 examines the role of green reverse logistics as a sustainable strategy towards circular economies in Africa. Chapter 11 explores the role of artificial intelligence in promoting green marketing practise. With the aid of a quasi-experiment, Chap. 12 examines the role of social media in disseminating green marketing messages. Chapter 13 provides the conclusion of this book by highlighting major insights gleaned from each chapter.

References

Agyabeng-Mensah, Y., Afum, E., & Ahenkorah, E. (2020). Exploring financial performance and green logistics management practices: examining the mediating influences of market, environmental and social performances. *Journal of cleaner production, 258*, 120613.

Aluko, O. A., Opoku, E. E. O., & Ibrahim, M. (2021). Investigating the environmental effect of globalization: Insights from selected industrialized countries. *Journal of Environmental Management, 281*, 111892.

Asongu, S. A., & Odhiambo, N. M. (2021). Enhancing governance for environmental sustainability in sub-Saharan Africa. *Energy Exploration & Exploitation, 39*(1), 444–463.

Bosch, S., & Schmidt, M. (2019). Is the post-fossil era necessarily post-capitalistic?—The robustness and capabilities of green capitalism. *Ecological Economics, 161*, 270–279.

Cruz, S. M. (2017). The relationships of political ideology and party affiliation with environmental concern: A meta-analysis. *Journal of Environmental Psychology, 53*, 81–91.

Gatti, L., Pizzetti, M., & Seele, P. (2021). Green lies and their effect on intention to invest. *Journal of Business Research, 127*, 228–240.

Jakob, M., Steckel, J. C., Jotzo, F., Sovacool, B. K., Cornelsen, L., Chandra, R., … & Urpelainen, J. (2020). The future of coal in a carbon-constrained climate. *Nature Climate Change, 10*(8), 704–707.

Jeannotte, M. S., & Dick, B. (2024). Measuring sustainability is hard to sustain: Lessons from a case study on collaborative cultural indicator development in Ottawa, Canada. *City, Culture and Society, 36*(100567).

Jha, P., & Schmidt, S. (2021). State of biofuel development in sub-Saharan Africa: How far sustainable?. *Renewable and Sustainable Energy Reviews, 150*, 111432.

Mendoza, M., Greenleaf, M., & Thomas, E. H. (2021). Green distributive politics: Legitimizing green capitalism and environmental protection in Latin America. *Geoforum, 126*, 1–12.

Mishra, D., Gunasekaran, A., Papadopoulos, T., & Hazen, B. (2017). Green supply chain performance measures: A review and bibliometric analysis. *Sustainable Production and Consumption*. https://doi.org/10.1016/j.spc.2017.01.003

Mkhize, S., & Ellis, D. (2024). Organic consumption as a means to achieve sustainable development goals and agenda 2063. *Sustainable Development*, 1–12.

Mondal, S., Singh, S., & Gupta, H. (2022). A meta-analysis of green and sustainable business models: A comprehensive approach. *Journal of Cleaner Production, 3711*(33623).

Mukonza, C., & Swarts, I. (2020). Te infuence of green marketing strategies on business performance and corporate image in the retail sector. *Business Strategy and the Environment, 29*(3), 838–845.

Namagembe, S., Ryan, S., & Sridharan, R. (2019). Green supply chain practice adoption and firm performance: manufacturing SMEs in Uganda. *Management of Environmental Quality: An International Journal, 30*(1), 5–35.

Sabela-Rikhotso, P. T. Z., van Niekerk, D., & Nemakonde, L. D. (2021). A critical analysis of the legal frameworks governing oil spill management in South Africa. *Marine Policy, 127*(104433).

Sakariyahu, R., Lawal, R., Etudaiye-Muhtar, O. F., & Ajide, F. M. (2023). Reflections on COP27: How do technological innovations and economic freedom affect environmental quality in Africa?. *Technological Forecasting and Social Change, 195*, 122782.

Seman, N. A. A., Govindan, K., Mardani, A., Zakuan, N., Saman, M. Z. M., Hooker, R. E., & Ozkul, S. (2019). The mediating effect of green innovation on the relationship between green supply chain management and environmental performance. *Journal of cleaner production, 229*, 115–127.

Schöggl, J. P., Rusch, M., Stumpf, L., & Baumgartner, R. J. (2023). Implementation of digital technologies for a circular economy and sustainability management in the manufacturing sector. *Sustainable Production and Consumption, 35*, 401–420.

Sharma, A. P. (2021). Consumers' purchase behaviour and green marketing: A synthesis, review and agenda. *International Journal of Consumer Studies, 45*(6), 1217–1238.
Sharma, N. K., & Kushwaha, G. S. (2019). Eco-labels: A tool for green marketing or just a blind mirror for consumers. *Electronic Green Journal, 1*(42).
Swinkels, L. (2024). Trading carbon credit tokens on the blockchain. *International Review of Economics and Finance, 91*, 720–733.
Terblanché-Greeff, A. C. (2019). *Ubuntu and environmental ethics: The West can learn from Africa when faced with climate change*, pp.93–109, African environmental ethics: A Critical Reader.
Wang, C., Zhang, Q., & Zhang, W. (2020). Corporate social responsibility, Green supply chain management and firm performance: The moderating role of big-data analytics capability. *Research in Transportation Business & Management, 37*, 100557.

CHAPTER 2

Beyond Formal Law: Harnessing African Indigenous Knowledge to Foster Green Marketing Governance and Sharing Economy

Chipo Mutongi and *Theo Tsokota*

Introduction

African countries are lagging behind in the race to attain sustainable development goals, including those linked to the natural environment (United Nations, 2021). As compared to other continents, the African continent is also regarded as more susceptible to climate change challenges (Terblanché-Greeff, 2019). The most common intervention by national governments

C. Mutongi (✉) • T. Tsokota
Department of Management Sciences, Faculty of Business Sciences, Midlands State University, Gweru, Zimbabwe
e-mail: mutongic@staff.msu.ac.zw; tsokotat@staff.msu.ac.zw

A. Muposhi, B. Nyagadza (eds.), *Sustainable Green Marketing Strategies for a Circular Economy*, Palgrave Studies of Marketing in Emerging Economies,
https://doi.org/10.1007/978-3-031-81627-7_2

to mitigate climate change drivers includes the enactment of environmental laws and implementation of international environmental policies. However, this approach has its limitations as the majority of African countries have limited resources that constrain effective enforcement of environmental laws and policies (Chibvongodze, 2016; Bishop, 2017). Moreover, environmental laws and policies have also been criticised for failing to consider the role of cultural values, customs and beliefs that are central to environmental management (Terblanché-Greeff, 2019). This view gains support from the Intergovernmental Panel on Climate Change (2012) which acknowledges the role of indigenous knowledge, culture, customs and beliefs in promoting green marketing governance. Similarly, Terblanché-Greeff (2019) notes that environmental values and beliefs central to the overarching African philosophy of Ubuntu are often not considered in the formulation and implementation of environmental policies.

This chapter therefore responds to increased calls to go beyond formal environmental laws and harness the potential power of African indigenous knowledge to foster green marketing governance. This chapter utilises Zimbabwe as a case study owing to its rich tapestry of traditions, culture, values and beliefs that have an impact on environmental governance. This chapter explores the potential of indigenous wisdom to inform and shape sustainable green marketing practices as well as unlocking the power of indigenous wisdom in leveraging traditional knowledge systems in driving sustainable green marketing practices to achieve a green economy. This chapter addresses the following objectives:

- To examine how indigenous knowledge can be leveraged to drive sustainable green marketing practices,
- To explore best practices in leveraging traditional knowledge to drive sustainable green marketing practices, and
- To assess the intersection of indigenous knowledge systems and green marketing governance.

Literature Review

African Traditions and Environmentalism

Africa's sustainable growth strategy acknowledges the role of indigenous knowledge systems such as customs, rituals and folklores in preserving the environment (Africa agenda, 2063). Indigenous knowledge systems are holistic practices, encompassing not just environmental stewardship but also social, cultural and spiritual dimensions (Berkes, 2012a). These systems have evolved over centuries, allowing indigenous people to thrive in diverse ecosystems. Moreover, indigenous communities often emphasise the importance of community over individualism, which can inform sustainable marketing strategies that prioritise a sharing economy and collective well-being rather than mere profit maximisation (Mason et al., 2015).

African indigenous wisdom is anchored in the ecocentric worldview that emphasises a holistic approach to living, recognising the interconnectedness of all living beings and the natural world (Berkes, 2012b; Patagonia, 2020). This worldview is reflected in traditional practices such as permaculture, agroforestry and eco-tourism, which prioritise sustainability and environmental stewardship (Patagonia, 2020).

African people are also known for a history of conserving biodiversity in the form of medicinal plants, including the use of traditional herbs in the production of products and services. This perspective helps businesses create products that resonate with consumers seeking authenticity and social responsibility. Makamure and Chimininge (2015) aver that African people believe that environmental protection and management are sanctioned by the creator, God and the ancestors of the land. African culture is centred on creating a harmonious relationship with spirits of the dead, animals, land and plant life. The way African people relate to land, animals, plant life and water bodies is regarded as environmentally conservative in nature. According to Makamure and Chimininge (2015), biospheric values are given more prominence than egoistic values that focus on self-interest and personal gain.

The Ubuntu Philosophy

The African philosophy of Ubuntu is centred on humanity, interconnectedness, compassion, community engagement and collective societal well-being (Kelbessa, 2015). It is underpinned by the principle that: 'I am because you are' (Tutu, 1999). This philosophy brings in the aspect of togetherness, team learning, knowledge sharing and belongingness. A key value proposed by the Ubuntu philosophy that is beneficial to the environment is sharing (O'Brien, 2009). This is because it guides against the value of materialism which is associated with acquisition of wealth for personal gain which leads to depletion of natural resources (Mugambiwa &

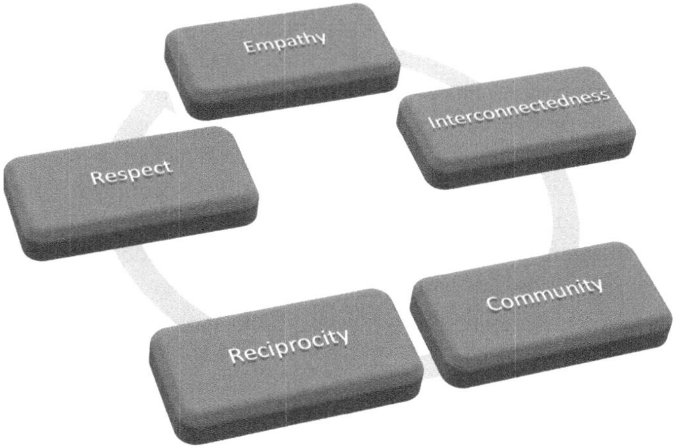

Fig. 2.1 Principles of Ubuntu

Rukema, 2019). Thus, the practice of the Ubuntu philosophy has the potential to assist in the development of green marketing strategies that promote sharing, collaboration and reduced consumption. According to Kelbessa (2015), green marketing governance may benefit from African principles of Ubuntu such as respect, empathy, interconnectedness, reciprocity and community. Figure 2.1 illustrates principles of Ubuntu:

Research Methodology

This chapter employs an interpretivistic research philosophy and a qualitative research method. It utilises interviews with key informants, reviews case studies and document analyses on indigenous knowledge as they relate to green marketing governance environmentalism with a focus on Zimbabwe.

Results and Discussion

Proverbs, Storytelling and Environmentalism

In African culture, proverbs are used to inculcate a sense of environmental responsibility, leading to the practise of green marketing (Chanza, 2015). The proverb *'aive madziva ava mazambuko'* can be translated to mean what was once a grand river is now a crossing place or ford. This proverb points to the fact that if our natural resources are not protected, it leads to the environmental degradation (Chimininge, 2015). The water in rivers can dry up and end up just having passageways. Another proverb which is used to promote environmentalism is: *'chenga ose manhanga hapana risina mbodzi'* translated to mean there is not a pumpkin without seeds, so value them all. This proverb stresses the importance of valuing everything in the ecosystem in order to maintain an intricate balance (Chimininge, 2015). *'similarly, Zano ndega akasiya jira mumasese'* which translates to mean people who forsake warnings and advice often fall into calamity is also uselful. This proverb is applied to environmental behaviour to warn people about the environmental challenges that are more likely to befall humankind if environmental sustainability is not prioritised. Another way to incorporate Indigenous wisdom in green marketing discourse is through storytelling (Chanza, 2015). Brands can share narratives that highlight their partnerships with indigenous communities or showcase traditional practices that align with sustainability. For example, companies like Patagonia have successfully integrated stories of Indigenous land stewardship into their marketing campaigns, thereby enhancing brand image while supporting environmental sustainability (Patagonia, 2020).

Indigenous Wisdom and Green Marketing Governance

The power of indigenous knowledge systems (IKS) can be harnessed to support green marketing governance and act as a community's armour against environmental shocks. Indigenous knowledge helps the communities to cope with periodic food shortages by utilising the traditional knowledge in preserving food and increasing food production which then ensures food security for all people. This is done through '*Zunde Ramambo*'. Zunde raMambo (the King's Granary) is a traditional social welfare system where a communal plot is farmed by the entire community for food security. This promotes the concept of Ubuntu including sustainable farming methods.

Taboos are also commonly used as an environmental management tool. Taboos are understood as specific rules that forbid people from performing certain actions (Chemhuru & Masaka, 2010). Taboos (*zviera*) form an essential part and parcel of the African morality. Among African people, environmental taboos play a pivotal moral role in the ontological well-being of both the person and the environment. Chemhuru and Masaka

(2010) add that prohibitions and restrictions through taboos on unsustainable use of certain plant species, forests, mountains, rivers, pools and non-human animals, among other ecological species in the ecosystem, are an established environmental sustainability practise. Taboo-oriented wisdom fosters a certain way of behaving that advances the well-being of the individual, the society and the environment. It is considered as a taboo to urinate in a river in Zimbabwean culture as it is assumed that it brings misfortune. This taboo promotes the practise of pro-environmental behaviour because urinating in the river negates a sound environmental ethic through pollution of water sources. Such behaviour also harms the aquatic creatures in the river as well as human beings who may need to use the water for domestic purposes. Through these taboos, the Zimbabwean culture has managed, over time, to ensure sustainable use of water resources.

In Zimbabwean communities, there is a heavy dependence on open wells, rivers and springs for water. One of the taboos that is used to preserve water bodies is *Ukachera mvura nechirongo chitema, tsime rinopwa*, which means ‘if you fetch water with a sooty black pot, the well will dry up’. It is important to note that this taboo discourages people from polluting water bodies. As Aschwanden (1989) notes, pools, springs and swamps sustain life. Therefore, such places are regarded as strategic resources that should be safeguarded. Also related to management of aquatic life, it is also considered as a taboo to kill a turtle (dafi) as it makes the well dry up. By observing this taboo, the rights of aquatic animals are preserved.

The Role Mermaids in the Preservation of Water Bodies

In Zimbabwean culture, mermaids are presumed to play a critical role in the preservation of water bodies. According to Mhaka (2015), the majority of *zvitubu* (water springs) found in Zimbabwe are believed to be sacred and habitats of mermaids. Fetching water from these springs using dirty containers or using metal cups and metal jars is strictly prohibited as these might cause the spring to dry up or cause the forbidden items to mysteriously disappear, or the culprit might be snatched by the water spirits. Some river pools are also said to be sacred, and humans are prohibited from washing their linen using soap or bathing in them using soap. Cutting down trees in and around sacred places is also prohibited. Mysterious punishment such as disappearance till appeasement is done or even death occurs in some areas if one cuts down trees or curses the forests. Sinking boreholes and construction of dams without consultation with njuzu and eventual appeasement is strictly prohibited in such spaces.

Our key informants also indicated that setting the forests around sacred water bodies on fire is customarily punishable by paying a fine in the form of a beast and home brew to appease the spirits. These sacred water springs are usually protected with logs and rocks or left open. People are only allowed to fetch water from these sources using *mukombe* (gourd) or a clay pot. Mhaka (2015) reports that women in their menstruation or those who have just given birth are not allowed anywhere near these sacred springs as they are seen as ritually unclean and are believed to cause the water to disappear.

Religion and Spirituality on Environmentalism Sacred Areas

The designation of some areas as sacred is another way of environmental preservation in Zimbabwe. Some of the sacred areas include Inyanga Mountains and Chinhoyi Caves. Sacred areas play a key role in biodiversity conservation. This long-lasting protection has allowed local communities to benefit from various services the forests provide, especially in times of dire need. The rules include not being allowed to cut down trees and not disturbing the rivers. For those who break the rules, history has it that they can disappear in the forests or rivers. These frightening consequences make people adhere to the traditional rules and regulations of the sacred places, hence leading to their preservation.

Values and Beliefs Related to Green Marketing Governance

The value orientation of collectivism dominates African culture and this translates to environmentalism (O'Brien, 2009). In particular, the Ubuntu philosophy espouses altruistic values that compel individuals to preserve the environment by fostering a group identity that emphasises societal benefit (Chanza, 2015). Environmental values have also evolved as communities have made several advances, such as mastering the positions of stars, the sun, and the moon to predict changes in weather patterns (Mugambiwa, 2017). All these experiences are crucial for anticipating future environmental changes and by explaining them within their cultural context. The spiritual world is also consulted through 'vadzimu' and 'tsvikiro' (Mugambiwa, 2017). If there is going to be a drought the Indigenous people brew indigenous beer called 'mukwerera' which they use to request for adequate rains. These practises have to be followed religiously such that the spirit mediums are not offended (Chanza, 2015).

Preservation of Indigenous Herbal and Organic Food

African practices also value the preservation of indigenous herbs and food in a manner that promotes biodiversity (Mugambiwa & Rukema, 2019). For example, the Rutaceae *Pterocarpus angolensis* and Aloe Vera are revered for their medicinal role. The largest zones of inhibition resulted from the plant extract Pterocarpus angolensis (Mubvamaropa root) against Escherichia coli. Some of the plants include *Clausena anisata*, Rutaceae (Muvhengahonye), used to treat diabetes, rheumatism, migraine headaches and worms, and the use of *Dicoma anamala*, Asteraceae (Chifumuro), used to treat stomach pain, colic, diarrhoea, syphilis, gonorrhoea, sore throats and coughs and colds. The fruits from indigenous trees such as baobab tree are also contributing to indigenous commercialised food, and there are increased calls for their preservation (Chanza, 2015; Mugambiwa & Rukema, 2019).

Implications of this Study

This chapter offers valuable insights on the role of indigenous knowledge in green marketing governance. Communities of practice should be called upon to share the Indigenous Knowledge Systems concerning green marketing. This can be done by prioritising environmental stewardship and social responsibility, incorporating traditional knowledge to develop green marketing governance policies, and collaborating with indigenous communities to develop sustainable green marketing strategies. Educating consumers about Indigenous wisdom and its relevance to sustainability is recommended. Campaigns that raise awareness about traditional ecological practices encourage consumers to appreciate the value of indigenous knowledge in addressing contemporary environmental challenges. This educational component can foster a deeper connection between consumers and brands committed to sustainability.

Conclusion

Unlocking the power of indigenous knowledge systems and wisdom offers a transformative pathway for sustainable green marketing practices. By embracing indigenous knowledge, businesses can enhance their authenticity, foster community engagement and contribute to environmental stewardship. However, this endeavour must be approached with respect and sensitivity to the rich cultural heritage of Indigenous people. As we move towards a more sustainable future, integrating Indigenous wisdom into green marketing not only benefits businesses but also honours the invaluable contributions of Indigenous communities to our understanding of sustainability. By harnessing the power of Indigenous wisdom, businesses can not only drive sustainable practices but also contribute to a more equitable and respectful world for all. Indigenous wisdom offers a wealth of knowledge and insights for driving sustainable green marketing practices. By embracing indigenous knowledge and environmental worldviews, green marketing can adopt a more holistic and long-term approach, prioritising ecological stewardship and sustainability.

References

Aschwanden, H. (1989). *Karanga mythology. An analysis of the conscious- ness of the Karanga in Zimbabwe*. Mambo Press.

Berkes, F. (2012a). *Sacred ecology: Traditional ecological knowledge and resource management*. Taylor & Francis.

Berkes, F. (2012b). *Sacred ecology*. Routledge.

Bishop, R. (2017). Confronting Climate Change Africa's Leadership on an Increasingly Urgent Issue. In Foresight *Africa: Top Priorities for the Continent in 2017*, ed. S. Amadou, 76–91. Brookings Institution.

Chanza, N. (2015). *Indigenous-based adaptation: An imperative for sustainable climate change strategies for Africa. Harnessing cultural capital for sustainability: A pan Africanist perspective*. Langaa Publishing House.

Chemhuru & Masaka. (2010). Taboos as sources of Shona People's environmental ethics. *Journal of Sustainable Development in Africa, 12*(7).

Chibvongodze, D. T. (2016). Ubuntu is not only about the human! An analysis of the role of African philosophy and ethics in environment management. *Journal of Human Ecology, 53*(2), 157–166.

Chimininge. (2015). *Going back to the roots: Tapping into African indigenous knowledge systems as a response to covid-19*. Retrieved September 28, 2024, from https://fis.uni-bamberg.de/bitstream/uniba/58422/6/fisba58422.pdf

Intergovernmental Panel on Climate Change [IPCC]. (2012). Glossary of Terms. In Managing the Risks of Extreme Events and Disasters to Advance Climate Change. A Special Report of Working Groups I and II of the Intergovernmental Panel on Climate Change (IPCC), ed. M. Tignor, and P.M. Midgley, 555–564. Cambridge, United Kingdom and New York, NY, USA, Cambridge University Press.

Kelbessa, W. (2015). Climate ethics and policy in Africa. *Thought and Practice: A Journal of the Philosophical Association of Kenya, 7*, 41–84.

Makamure and Chimininge. (2015). *International Journal of Humanities and Social Science Invention., 4*(11), 07–12.

Mason, J., et al. (2015). The three sisters: A traditional agricultural practice. *American Indian Culture and Research Journal.*

Mhaka, E. (2015). Sustainable agriculture and culture: An exploration of the significance of the indigenous Shona religious beliefs and the link between indigenous knowledge systems practices in agricultural sustainability. *Sch. J. Arts Humanit. Soc. Sci., 3*(1E), 298–230.

Mugambiwa, S. S. (2017). Knowledge of climate change and the use of indigenous practice to adapt to climate hazards in Mutoko Rural Community in Mashonaland East Province, Zimbabwe (Doctoral dissertation).

Mugambiwa, S. S., & Rukema, J. R. (2019). Rethinking indigenous climate governance through climate change and variability discourse by a Zimbabwean rural community. *International Journal of Climate Change Strategies and Management, 11*(5), 730–743.

O'Brien, K. (2009). Values and the Limits to Adaptation. In *Adapting to Climate Change: Thresholds, Values, Governance*, ed. W.N. Adger, I. Lorenzoni and K.L. O'Brien. Cambridge University Press.

Patagonia. (2020). *Our commitment to indigenous rights.* [Patagonia website]. https://www.patagonia.com

Terblanché-Greeff, A. C. (2019). Ubuntu and environmental ethics: The west can learn from Africa when faced with climate change. In M. Chemhuru (Ed.), *African environmental ethics. The international library of environmental, agricultural and food ethics* (Vol. 29). Springer.

Tutu, D. (1999). No Future Without Foregiveness. London: Rider Random House.

United Nations. (2021). The Sustainable Development Goal Report (2021) https://unstats.un.org/sdgs/report/2021/The-Sustainable-Development-Goals-Report-2021. (Accessed 20 April 2024).

CHAPTER 3

Green Corporate Governance and Sustainable Development Goals: Lessons for Africa from Global North

Mufaro Dzingirai, Kudakwashe Zvitambo, and Herbert Masukume

Introduction

In view of the growing concerns related to environmental sustainability issues, it is prudent for the board members to have a deeper understanding of how companies are governed with special emphasis on green marketing and sustainability. Green corporate governance refers to the process of governing and directing the company to satisfy all the stakeholders while striking a balance between human interests and the natural environment (Usman et al., 2024; Wang et al., 2023). At corporate level, green marketing tenets

M. Dzingirai (✉) • K. Zvitambo • H. Masukume
Department of Management Sciences, Faculty of Business Sciences, Midlands State University, Gweru, Zimbabwe
e-mail: dzingiraim@staff.msu.ac.zw; zvitambok@staff.msu.ac.zw; masukumeh@staff.msu.ac.zw

A. Muposhi, B. Nyagadza (eds.), *Sustainable Green Marketing Strategies for a Circular Economy*, Palgrave Studies of Marketing in Emerging Economies,
https://doi.org/10.1007/978-3-031-81627-7_3

and reporting standards are captured in the International Financial Reporting Standards (IFRS) S1 and S2. Consequently, the board members have to formulate a green governance strategy in order to address the pressing issues related to sustainability. Effective green corporate governance strategies play a crucial role in creating responsible business values. With the pressing issues of climate change and Sustainable Development Development Goals (SDGs), responsible businesses are pressured to ensure environmental protection through various initiatives that deal with environmental degradation and climate change (Khan et al., 2022; Mutanda et al., 2024). The concept of sustainability reporting is also advocated for as a way of fostering green corporate governance (Dinçer et al., 2023; Wang et al., 2023).

While green marketing is considered to be a corporate board issue, it appears that board members are lagging behind in terms of how to create a vision statement that encompasses sustainability issues (Dinçer et al., 2023). This has cascaded down to skewed mission statements, goals, objectives, and values. Sustainability reporting is at the epicentre of green corporate governance as it supports disclosure and transparency when it comes to the interface between businesses and the natural environment. This suggests that companies are required to disclose sustainability-related indicators and metrics related to sustainability risks and opportunities. Consequently, the international sustainability standards and jurisdiction-based indicators act as the guidelines in the formulation of the most effective green corporate governance strategies.

In the context of Africa, companies listed on the stock exchange are mandated to adopt sustainability and Environmental, Social, and Governance (ESG) reporting in line with the promotion of green governance. This is a positive development towards mandatory green governance in Africa. This could go a long way in reducing environmental degradation.

Although African economies adopted SDGs and committed to international environmental treaties such as Net zero targets, the integration of green marketing practice at corporate level remains a major challenge (Makpotche et al., 2024). Scholars who investigated green corporate governance concentrated on green innovation and green financing (Makpotche et al., 2024; Wang et al., 2024). Hence, the literature on green corporate governance needs further attention. To bridge these knowledge gaps, this study adopted a comprehensive and systematic research methodology.

This chapter seeks to address the following objectives: the first section examines the interconnectedness of green corporate governance and sustainable development goals, and to recommend effective strategies for fostering green marketing governance. The remainder of this chapter is structured as follows: The second section covers a critical review of the stakeholder theory as a theoretical framework, green corporate governance, the nexus of green corporate governance and SDGs, and trends in green corporate governance. The third section covers the methodology that underpins this investigation. The fourth section reports the research findings and discussion. The fifth section is dedicated to the implications of the study.

Literature Review

Stakeholder Theory

The stakeholder theory enables the company to identify all the key stakeholders who affect or are affected by the company's operations and decisions (Pies & Valentinov, 2024). Literature shows that the company should actively engage and solicit input from a wide range of stakeholders, not just shareholders (Gomes et al., 2024). The company should strive to find fair, and equitable ways to balance and reconcile competing interests. In the context of green corporate governance, stakeholder theory helps researchers to understand and implement green corporate governance (Awa et al., 2024). Literature indicates that stakeholder theory encourages companies to identify all the relevant stakeholders affected by their environmental and sustainability practices, including employees, customers, suppliers, local communities, and environmental advocacy groups (Valentinov & Iliopoulos, 2024). Furthermore, stakeholder theory recognizes that different stakeholder groups have conflicting interests when it comes to environmental and sustainability issues. Moreover, the stakeholder theory emphasizes the importance of actively engaging with stakeholders, seeking their input, and incorporating their perspectives into decision-making (Sanusi & Assih, 2024). It calls for increased transparency and accountability around a company's environmental and sustainability performance. According to Wang and Feng (2024), the theory encourages companies to adopt a long-term, sustainable perspective, rather than focusing only on short-term profits. Therefore, stakeholder theory provides the framework for systematically managing relationships with different stakeholder groups as part of green corporate governance (Lei & Yu, 2024; Salihi et al., 2024). This helps companies to navigate the complex environmental and sustainability trade-offs.

Conceptualization of Green Corporate Governance

There is a lack of consensus in the extant literature on the definition of green governance (Allie & Sudibijo, 2024; Puspitasari et al., 2024; Zhang et al., 2024). Other scholars describe green governance as the act of integrating environmental and sustainability considerations into a company's governance structure and decision-making processes (Allie & Sudibijo, 2024; Zhang et al., 2024). Another school of thought describes green governance as directing and controlling entities towards the sustainability of environmental sustainability (Puspitasari et al., 2024). These two views differ in that: the first assumes that green corporate governance involves making decisions on environmental issues, whilst the second regards environmental issues as an aspect discussed at the board level of an entity. Failure by scholars to agree on the meaning of green governance creates a vacuum in its understanding. In this chapter, green corporate governance is conceptualised as the process of governing and directing the company to satisfy all the stakeholders while striking a balance between human interests and environmental well-being.

Literature indicates that companies establish clear environmental policies, set measurable sustainability goals, and integrate these into their overall business strategy (Purbawati & Mujiyati, 2024). However, information on the implementation and funding of green corporate governance remains unclear. Adding to the discourse, Pasha et al. (2024) argue that the board of directors should take an active role in overseeing the company's environmental and sustainability performance, often through dedicated board committees. Nevertheless, the structure of many boards does not reflect such arrangements, making it difficult to implement. On the other hand, Yang et al. (2024) postulate that companies are showing commitment in terms of reporting their environmental impact, resource usage, emissions, waste management, and other sustainability metrics in their annual reports or dedicated sustainability reports. According to these scholars, it indicates a degree of implementing green corporate governance. Sanusi and Assih (2024) suggest that companies should tie a portion of executive compensation to the achievement of environmental and sustainability targets, incentivizing leadership to prioritize these issues.

Similarly, Wang and Feng (2024) suggest that companies should engage with shareholders, customers, employees, and local communities to understand their environmental concerns and incorporate feedback into decision-making. However, companies in developing countries face a plethora of challenges that lead them to concentrate on survival rather than other issues. In addition, Lei and Yu (2024) recommend that companies should evaluate the environmental practices of their suppliers and work collaboratively to improve the sustainability of their supply chains. This only holds water when the company has bargaining power over the suppliers, but in the current situation where the majority of companies in Africa import almost every raw material, it becomes difficult. Contributing to the discussion, Salihi et al. (2024) suggest that companies invest in and deploy technologies that reduce their carbon footprint, improve energy efficiency, and minimize their environmental impact. Although the idea sounds noble, scarcity of literature on this aspect in African corporate literature is worrying.

Literature reveals that the goal of green corporate governance is to embed environmental considerations into the core of a company's operations and decision-making, align business practices with sustainability principles, and drive long-term value creation while mitigating environmental risks (Allie & Sudibijo, 2024; Sanusi & Assih, 2024). However, the limited availability of scholarly literature on green corporate governance in Africa indicates the lack of research and development in companies, suggesting that the boards do not consider the issue of green corporate governance.

Green Governance and Sustainable Development Goals

Searching using various engines, including Google, indicates that the nexus of green corporate governance and sustainable development goals remains under-researched. Many journal articles indicate studies on sustainable development goals (Almaqtari et al., 2024; Behera et al., 2024; Lyulyov et al., 2024b; Usman et al., 2024; Phan, 2024; Saha et al., 2024) and a few on green corporate governance (Ahmed & Anifowose, 2024; Hoang et al., 2024; Makpotche et al., 2024; Pamungkas et al., 2024; Patnaik, 2024; Yang et al., 2024). Literature indicates that those scholars who tried to link green corporate governance to sustainable development goals only picked a few SDGs (Casciello et al., 2024; Saha et al., 2024; Sharma & McLean, 2024). However, these scholars did not link sustainable development goals to the board functions. The role of the board of directors in implementing green corporate governance to achieve sustainable development goals, especially in developing countries remains unclear.

According to Sungkawati (2024), only a few sustainable development goals were captured in the discussion of green governance. The scholar reports that SDG 6 on clean water and sanitation ensures universal and equitable access to safe and affordable drinking water for all. Sungkawati postulates that green corporate governance measures around water usage, pollution prevention, and wastewater management can contribute to SDG 6. However, the literature does not clearly state the green corporate governance implementation matrix and the role of board members in ensuring the achievement of SDG 6.

Similarly, Lyulyow et al. (2024b) highlighted the issues related to SDG 7 which is affordable and clean energy. The scholars explained that green corporate governance encourages companies to prioritize the use of renewable energy sources, such as solar, wind, and hydropower, in their operations. In support, Pathiranage et al. (2024) posited that green corporate governance frameworks mandate the implementation of robust energy efficiency measures across a company's facilities, processes, and product design. However, the scholars do not provide the framework for the nexus of green corporate governance and sustainable development goals.

Sitorus and Sinaga (2024) explained SDG 12, which is related to responsible consumption and production as it relates to green corporate governance. The scholars argue that green corporate governance focuses on sustainable procurement, waste reduction, and circular economy principles that directly support SDG 12.

Literature shows that scholars believe that the implementation of green corporate governance enhances the achievement of SDGs (Yanto et al., 2024). However, they do not provide the green corporate governance framework and the implementation matrix. Furthermore, their explanation failed to reveal the funding model when implementing the green corporate governance framework.

Literature indicates that green corporate governance implementation leads to the achievement of SDG 13, which is climate action (Casciello et al., 2024). According to Makpotche et al. (2024), a green corporate governance framework requires companies to set ambitious targets for reducing their emissions, aligning with the goals of SDG 13 to take urgent action to combat climate change and its impacts. This includes implementing strategies to improve energy efficiency, shift to renewable energy sources, and optimize operational processes to minimize carbon footprint. However, these scholars do not provide the implementation matrix that helps companies in developing countries.

Similarly, Sharma and Mclean (2024) explained the implementation of green corporate governance to achieve SDG 14, which is life below water. The scholars explained that a green corporate governance framework encourages companies to implement robust waste management systems and adopt circular economy principles to prevent and substantially reduce marine pollution, particularly from land-based activities. In addition, Patnaik (2024) postulates that a green corporate governance framework promotes the responsible and sustainable sourcing of marine-based resources, ensuring that companies' supply chains do not contribute to the overexploitation of marine ecosystems. However, the scholars conducted studies in countries close to the oceans and left out landlocked countries. The application of their recommendations falls short for these landlocked countries. Furthermore, many studies concentrated in developed countries where companies do not struggle to survive, hence the availability of funds for green corporate governance implementation in achieving sustainable development goals.

Literature indicates that green corporate governance can be implemented on SDG 15 which is life on land, which includes biodiversity protection, sustainable land use, and forestry practices (Yang et al., 2024). Studies show that the green governance framework requires companies to assess and mitigate their impacts on terrestrial biodiversity, supporting the conservation, restoration, and sustainable use of ecosystems as outlined in SDG 15 (Sharma & McLean, 2024). This includes measures to protect and prevent the loss of forests, wetlands, and other critical habitats. In support, Pathiranage et al. (2024) argue that green governance encourages companies to adopt sustainable land use practices that minimize deforestation, land degradation, and habitat fragmentation, in line with SDG 15's targets.

In addition, scholars explain the implementation of green corporate governance in achieving SDG 17, which focuses on partnerships for the goals. Phan (2024) states that a green corporate governance framework promotes the establishment of multi-stakeholder partnerships and collaborative initiatives involving companies, governments, civil society, and other stakeholders to collectively address sustainability challenges.

However, the scholars do not provide the framework for the nexus of green corporate governance and sustainable development goals. Contributing to the discourse, Sharma and McLean (2024) report that green corporate governance fosters the sharing of best practices, technological innovations, and expertise among companies, governments, and other stakeholders to enhance global capacity for sustainable development. This aligns with SDG 17's targets of enhancing international support for implementing effective and targeted capacity-building in developing countries. Moreover, green corporate governance encourages companies to actively engage with policymakers and contribute to the development of coherent and aligned policies and regulations that support the achievement of the SDGs. Nevertheless, literature on the nexus of green corporate governance and sustainable development goals selects certain SDGs where to apply them instead of taking a holistic approach.

Trends in Green Corporate Governance

According to Shin et al. (2022), green corporate governance is described as a combination of environmental reflections in the decision-making process of organizations for the purposes of improving sustainability as well as reducing negative effects that affect the environment. In the same view as Shin et al. (2022) were Yu and Choi (2023) who posited that the green corporate governance concept has become a major concern for most organizations due to issues related to climate change, environmental footprints, and resource depletion. Pham et al. (2021) also stated that organizations in the twenty-first century are not putting in place measures to increase awareness of green corporate governance to improve sustainability growth with regard to meeting long-term objectives. In support of Pham et al. (2021) were Liao et al. (2023) who stated that green corporate governance is vital for organizations as it helps in the preservation of resources that can be used to improve sustainable development and preservation of the environment. A study conducted by Grewal et al. (2017) made suggestions that working with shareholders can improve an organization's ability to improve environmental awareness and adoption of greener practices that influence overall sustainability performance.

However, Li and Tang (2022) stated that a number of studies have been conducted on trends in green corporate governance, and they proved that most of them dwelled on disclosure as well as sustainability reporting. Numerous companies are now aware of green corporate governance, and they are publishing reports related to the concept, showing its influence on social responsibility practices (Kim & Kim, 2021). Publishing detailed green corporate governance is essential in showing the trends as well as promoting transparency to all the shareholders in the business domain and shows the level of commitment to the formulation of sustainability strategy (Jo & Harjoto, 2020). A study conducted by Klynveld Peat Marwick Goerdeler (KPMG) (2017) on trends in green corporate governance showed that there is a significant 75% increase in organizations across the global village that are publishing reports on sustainability. The trend shows that organizations are recognizing the essential part of showing a level of commitment with regard to sustainability. A similar trend in green corporate governance illustrates the organization's ability to implement environmental management systems that play a key role in mitigating as well as identifying indicators of environmental impact that include waste management reduction, pollution control, and energy conservation (Hasan et al., 2020).

Green Corporate Governance and Corporate Sustainability

According to Goss et al. (2022), there is a positive relationship between green corporate governance and corporate sustainability. In support of Goss et al. (2022) were Karna et al. (2023) who stated that green corporate governance promotes corporate sustainability in the view that organizations that practice green corporate governance have the ability to increase awareness among all stakeholders, preserve the environment as well as reduce depletion of resources which can be used to improve corporate sustainability in the long run. Notably, green corporate governance is vital in promoting corporate sustainability through the view that organizations will be able to build trust from other stakeholders when they provide detailed, timely publication reports on green corporate governance (Flammer, 2021). A study conducted by El Ghoul et al. (2021) on the relationship between green corporate governance and corporate sustainability showed that organizational value can be achieved through engaging in corporate social responsibility at the country level. According to Shrivastava and Addas's (2023) research, improved green corporate governance is associated with improved sustainability performance and a higher likelihood of businesses adopting sustainability policies and adhering to international sustainability standards.

In contrast, Cheung and Zhang (2022) stated that in as much as green corporate governance promotes corporate sustainability, other determinants need to be taken into consideration, for instance, existence of uncontrollable factors and leadership competencies. In the same view, Liao et al. (2023) stated that green corporate governance can be hard to practice in the organization if the leader is corrupt or does not involve the stakeholders in the decision-making process, hence affecting corporate sustainability. According to Bai et al. (2019), uncontrollable factors such as political instability, economic instability as well as natural disasters have a negative impact on the success of green corporate governance hence affecting corporate sustainability. A study conducted by Adams and Mackenzie (2020) showed an inverse relationship between green corporate governance and corporate sustainability as the study pointed out the issue of availability of resources propelling suitability performance in firms. Good governance should make sure that these limited resources are used effectively to meet the demands of sustainable development, given the growing needs and resource constraints (Aguilera et al., 2020).

Methodology

This study on green corporate governance adopted a systematic literature review methodology that is hailed for providing a comprehensive understanding of a particular research phenomenon (Dzingirai & Baporikar, 2022). To foster the credibility and trustworthiness of this methodology, the Preferred Reporting Items for Systematic reviews and Meta-Analyses (PRISMA) approach to systematic review of existing literature was deemed appropriate for this study. The PRISMA approach is widely accepted as the robust way to identify, assess, map, and visualize the emerging patterns from the current literature (Bejarano et al., 2023; Mhlanga & Dzingirai, 2024; Usman et al., 2024). In line with the PRISMA approach, the search strategy was informed by the following search terms: green corporate governance, green corporate strategy, and green corporate governance implementation. The search strategy was limited to peer-reviewed journal articles in the fields of business, strategic management, and accounting fields. Book chapters, conference papers, reviews, books, conference reviews, erratum, editorials, and data papers were excluded. The search yields 274 usable articles. The outcome of the search strategy is presented below (Fig. 3.1):

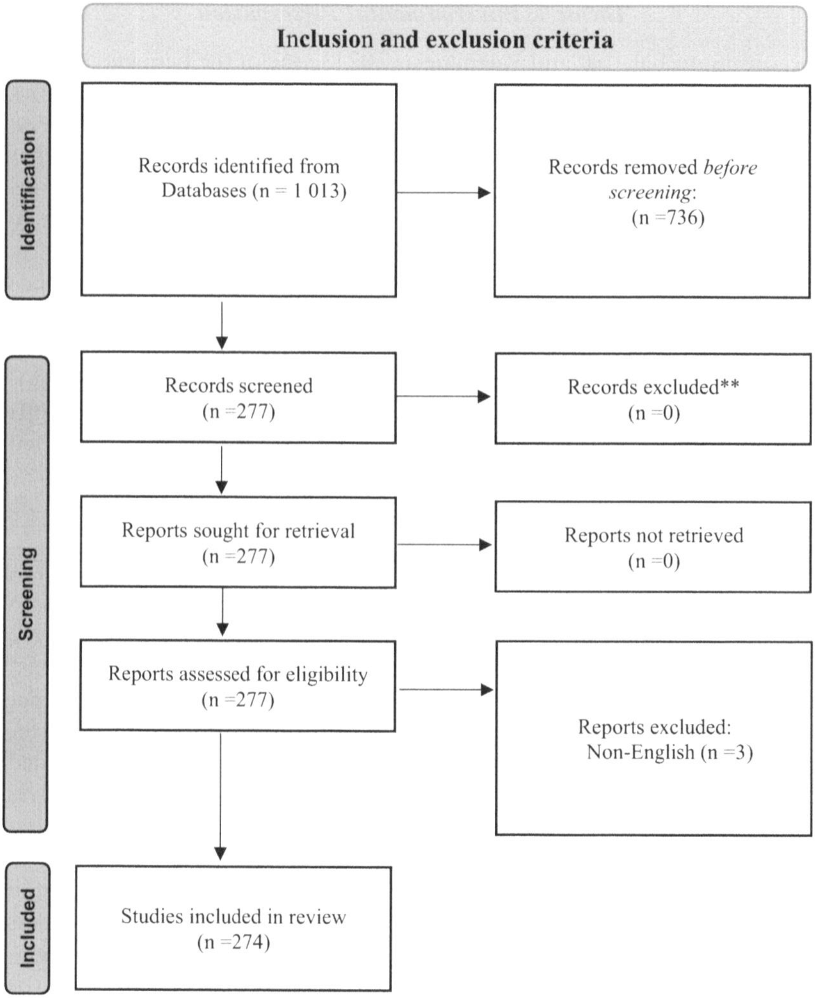

Fig. 3.1 PRISMA flowchart

Results and Discussion

This section is dedicated to the results and discussion of this study on green governance in an attempt to provide a comprehensive understanding of how board members can benefit from the disclosure of materiality information related to sustainability. The following results are reported and discussed as per themes that emerged:

Theme 1: Environmental Preservation

According to Balatbat and Valenzuela (2021), one of the benefits of green corporate governance is environmental preservation. Organizations work hand in hand with the environment through the view that they access raw materials from the environment and they must put in place policies that are eco-friendly (Chatterji & Toffel, 2019). Operations of organizations pose dangers to the environment with regards to pollution, climate change, increased environmental footprints, and land degradation; hence, it is essential for companies to put in place green corporate governance practices that improve environmental preservation (Dey et al., 2019). The total defence of a natural area against human activity is known as environmental preservation. The distinction between preservation and conservation is that the former limits human involvement, while the latter prohibits all commercial, industrial, and development activities within a defined region (Pham et al., 2021).

Theme 2: Build Stakeholder Trust

According to Du et al. (2020), the other benefit of green corporate governance is promoting stakeholder trust. Stakeholders in business are important to the success of any business, and if their trust is affected, then this can affect organizational success (El Ghoul et al., 2021). Stakeholder trust can be achieved through a company's ability to publish detailed, trustworthy reports on green corporate governance unlike what happened in the case of Enron (Flammer, 2021). Houston, Texas-based Enron was a utility and energy trading corporation that committed one of the worst accounting frauds in history. Enron's leaders used accounting tricks to inflate the company's sales, making it the seventh-largest corporation in the United States for a while. After the scam was discovered, the business collapsed fast, and in December 2001, it filed for Chapter 11 bankruptcy. Enron was secretly creating false financial records and exaggerating the company's achievements prior to being discovered. Despite the fact that the organization did see operational success in the 1990s, the company's wrongdoings were ultimately made public in 2001 (Goss et al., 2022).

Theme 3: Transparency and Accountability

According to Hasan et al. (2020), the other benefit of green corporate governance is promoting transparency and accountability. Organizations need to try by all means to be as transparent and accountable as possible when it comes to promoting green corporate governance, as this can improve environmental sustainability (Jo & Harjoto, 2020). Management needs to be accountable for their actions in the workplace by providing stakeholders with green corporate governance information that is not fabricated (Kang & Moon, 2019). Organizations also need to be accountable for their actions, ensuring that they are held accountable for their operational actions that may be in line with pollution and destruction of the environment (Karna et al., 2023). A company's commitment to green corporate governance can be held accountable by stakeholders through transparency. Stakeholders can evaluate whether a firm is fulfilling its commitments and take necessary corrective action when it discloses information about its initiatives and advancements in an open and transparent manner (Kim & Kim, 2021). Since it gives stakeholders a chance to voice their concerns about the environment, offer feedback, and have meaningful interactions with the organization, transparent communication promotes dialogue between businesses and stakeholders (Li & Tang, 2022).

Theme 4: Compliance and Regulatory Advantages

Organizations that uphold green corporate governance principles are more likely to conform to changing environmental laws as well as adhere to guidelines (Shin et al., 2022). Businesses can avoid costly fines, compromised business reputation, and pollution lawsuits by adhering to environmental standards. Financial rewards for compliance, such as grants, tax exemptions, and subsidies, can encourage sustainable behaviour (Yu & Choi, 2023). This lowers legal risks while also showing a dedication to ethical corporate conduct that can improve ties with authorities. According to Garcia and Lee (2022), by ensuring that firms follow environmental standards, compliance monitoring helps to protect the environment and conserve natural resources. The increased attention that green movements have received in recent years has increased the pressure on businesses to address environmental issues (Chen & Wang, 2023).

Theme 5: Promote Green Innovation

It emerged that green governance is very important in the sense that it promotes green innovation, whereby the organization strives to address environmental concerns emanating from business activities (Chen & Wang, 2023). It is a fact that business activities can pose serious environmental problems. For instance, manufacturing firms can accelerate water, air, and land pollution, which can negatively affect our environment. To make matters worse, manufacturing firms emit greenhouse gases that can cause ozone layer depletion. This can lead to global warming. As such, a company's effort to reduce greenhouse gas emissions can embark on green innovations like the use of clean energy. Clean energy can be in the form of solar, wind, and gas energy (Kim & Kim, 2021).

Implications of the Study

The theoretical, practical, and policy implications can be derived from this study on the interaction between green governance and SDGs. Theoretically, this study adds value by extending the knowledge boundaries related to green corporate governance by taking stock of the existing body of knowledge. Concerning practical implications, the board of directors and executive management can gain an in-depth understanding of how green corporate governance can add value to IFRS S1 & S2 and sustainability reporting. Moreover, educators and trainers can incorporate the aspect of green corporate governance into their management curriculum, given that in many countries sustainability reporting is now mandatory for publicly listed companies in Africa. Finally, there are some policy implications linked to this study. In this regard, government agencies can legalize key tenets of green corporate governance, like forcing companies to report how they are causing environmental risks and opening environmental opportunities.

Going forward, the upper echelons that is board members are expected to embrace green leadership practices whereby they promote environmentally friendly practices in the organization. This can be in various forms like compensation strategy, green innovation, strategy, and green policy (Dinçer et al., 2023; Wang et al., 2023). To inculcate a green culture, the board members mainly adopt the 6Rs of sustainability, namely, rethink, reuse, recycle, repair, refuse, and reduce. These 6Rs feed into the overall green corporate governance strategy and policy. More interestingly, stakeholders like investors are interested in green-oriented organizations which implies that companies are now expected to disclose environment-related risks and opportunities in a transparent manner so that they make informed decisions. According to IFRS SI and S2, companies can disclose material information pertaining to sustainability issues. Therefore, it is imperative to note that an effective green corporate governance strategy acts as a powerful tool to report climate-related and environmental-based risks and opportunities (Khan et al., 2022; Usman et al., 2024). Notably, the issues like water consumed, electricity consumed, and greenhouse gases emitted must be disclosed in the annual report.

Conclusion

This chapter has covered the interconnectedness of green corporate governance and SGDs, taking a systematic review approach. With the increasing attention to sustainable development goals and corporate sustainability, it is emerging that green corporate governance increasingly becomes more relevant than before. As such, the issues of fairness, sustainability reporting, transparency, and disclosure are at the epicentre of green corporate governance. Surprisingly, little attention has been paid to green corporate governance in the extant governance discourse and scholarship. Keeping this in mind, this chapter established that companies that embark on green corporate governance can ensure environmental preservation, build stakeholder trust, promote transparency and accountability, foster compliance, and promote green innovation. It is concluded that board diversity can also be assessed in terms of green corporate governance knowledge in an effort to promote adherence to sustainability reporting standards captured in IFRS S1 and S2. This study could assist practitioners and policymakers.

References

Adams, R. B., & Mackenzie, C. (2020). Corporate social responsibility and firm risk: Theory and empirical Evidence. *Annual Review of Financial Economics, 12*, 67–87.

Aguilera, R. V., Desender, K. A., Bednar, M. K., & Lee, J. H. (2020). Connecting the dots: Bringing external corporate governance into corporate governance research. *Academy of Management Annals, 14*(1), 1–41.

Ahmed, A., & Anifowose, M. (2024). Corruption, corporate governance, and sustainable development goals in Africa. *Corporate Governance: The International Journal of Business in Society, 24*(1), 119–138.

Allie, R., & Sudibijo, Y. A. (2024). Analysis of the effect of green banking, corporate social responsibility, and ultimate ownership on bank performance with good corporate governance as a moderating variable. *Technium Sustainability, 5*, 75–89.

Almaqtari, F. A., Elsheikh, T., Hussainey, K., & Al-Bukhrani, M. A. (2024). Country-level governance and sustainable development goals: Implications for firms' sustainability performance. *Studies in Economics and Finance, 41*(3), 684–723.

Awa, H. O., Etim, W., & Ogbonda, E. (2024). Stakeholders, stakeholder theory and Corporate Social Responsibility (CSR). *International Journal of Corporate Social Responsibility, 9*(1), 11.

Bai, C., Rao, Y., & Shen, Y. (2019). Greenwashing and green governance: Evidence from Chinese firms' pollution disclosures. *Journal of Business Ethics, 160*(3), 751–769.

Balatbat, M. C. A., & Valenzuela, R. L. (2021). Corporate environmental responsibility and financial performance: The moderating role of ownership concentration. *Asia-Pacific Journal of Accounting and Economics, 28*(3), 397–417.

Behera, B., Behera, P., & Sethi, N. (2024). Decoupling the role of renewable energy, green finance, and political stability in achieving the sustainable development goal 13: Empirical insight from emerging economies. *Sustainable Development, 32*(1), 119–137.

Bejarano, J. B. P., Sossa, J. W. Z., Ocampo-López, C., & Ramírez-Carmona, M. (2023). Open innovation: A technology transfer alternative from universities. A systematic literature review. *Journal of Open Innovation: Technology, Market, and Complexity, 9*, 100090.

Casciello, R., Maffei, M., & Meucci, F. (2024). Board characteristics and Sustainable Development Goals disclosure: Evidence from European state-owned enterprises. *Journal of Public Budgeting, Accounting & Financial Management*. Vol. ahead-of-print No. ahead-of-print.

Chatterji, A. K., & Toffel, M. W. (2019). The role of environmental and social performance in corporate governance: A research agenda. *Organization Science, 30*(1), 1–20.

Chen, L., & Wang, Q. (2023). Green corporate governance and firm performance: A meta-analysis. *Journal of Business Ethics, 41*(1), 78–94.

Cheung, Y. L., & Zhang, H. (2022). Do corporate governance mechanisms influence environmental disclosures? Evidence from China. *Journal of Business Ethics, 175*(1), 19–37.

Dey, A., Hasan, I., & Hossain, M. (2019). Corporate governance and green innovation: International evidence. *Journal of Business Ethics, 155*(2), 401–435.

Dinçer, H., Kalkavan, H., Yüksel, S., & Karakuş, H. (2023). Defining the most effective green corporate governance strategies for sustainable performance. In *Economic, environmental and health consequences of conservation capital: A global perspective* (pp. 41–54). Singapore.

Du, X., Hou, Q., & Wang, J. (2020). Does corporate governance affect corporate environmental information disclosure? Evidence from China. *Journal of Business Ethics, 166*(2), 293–310.

Dzingirai, M., & Baporikar, N. (2022). Trends and patterns in Turnaround strategies: A bibliometric perspective and Future Research Agenda. *International Journal of Sociotechnology and Knowledge Development, 14*(1), 1–15.

El Ghoul, S., Guedhami, O., & Kim, H. (2021). Country-level institutions, firm value, and the role of corporate social responsibility initiatives. *Journal of International Business Studies, 52*(7), 1172–1199.

Flammer, C. (2021). Corporate governance and the rise of integrating corporate social responsibility criteria in executive compensation: Effectiveness and implications for firm outcomes. *Academy of Management Journal, 64*(1), 116–147.

Garcia, M., & Lee, S. (2022). The role of board diversity in green corporate governance: Evidence from the US. *Journal of Environmental Management, 28*(4), 112–129.

Gomes, R. C., Lisboa, E., Sarturi, G., & Mirapalheta, G. (2024). How has stakeholder theory served the public administration literature? In search of the intellectual structure of the field. *Public Management Review*, 1–23.

Goss, A., Roberts, G. S., Srinivasan, S., & Wu, Y. J. (2022). The impact of corporate social responsibility on risk taking and firm value in the energy industry. *Journal of Business Ethics, 182*(3), 663–682.

Hasan, I., Kobeissi, N., & Zhang, H. (2020). Corporate social responsibility and firm financial performance: The mediating role of productivity. *Journal of Business Ethics, 165*(4), 625–642.

Hoang, P. D., Nguyen, L. T., Tran, B. Q., & Ta, D. T. (2024). Corporate governance for sustainable development in Vietnam: Criteria for SOEs based on MCDM approach. *Plos one, 19*(5), e0302306.

Jo, H., & Harjoto, M. A. (2020). Corporate governance and firm value: The impact of corporate social responsibility. *Journal of Business Ethics, 141*(2), 337–351.

Kang, H., & Moon, J. (2019). The Relationship between Corporate Governance and Corporate Social Responsibility in Asia: Evidence from Korea and Japan. *Journal of Business Ethics, 160*(3), 757–778.

Karna, A., Hansen, E. G., & Wocken, C. (2023). Corporate social responsibility and stock returns: evidence from the european banking sector. *Journal of Banking & Finance, 136.*

Khan, P. A., Johl, S. K., & Akhtar, S. (2022). Vinculum of sustainable development goal practices and firms' financial performance: A moderation role of green innovation. *Journal of Risk and Financial Management, 15*(3), 96.

Kim, Y., & Kim, S. (2021). The influence of corporate governance on sustainable corporate social responsibility activities: Evidence from Korea. *Journal of Business Ethics, 168*(4), 771–789.

Lei, X., & Yu, J. (2024). Striving for sustainable development: Green financial policy, institutional investors, and corporate ESG performance. *Corporate Social Responsibility and Environmental Management, 31*(2), 1177–1202.
Li, J., & Tang, Y. (2022). Corporate social responsibility and firm value: A meta-analytic review. *Journal of Business Ethics, 175*(3), 487–505.
Liao, L., Zhou, L., & Wang, Y. (2023). Board gender diversity and corporate social responsibility: Evidence from China. *Journal of Business Ethics, 171*(2), 239–254.
Lyulyov, O., Pimonenko, T., Chen, Y., Kwilinski, A., & Yana, U. (2024a). Countries' green brands within the context of sustainable development goals. *Journal of Innovation and Knowledge, 9*(3), 100509.
Lyulyov, O., Pimonenko, T., Saura, J. R., & Barbosa, B. (2024b). How do e-governance and e-business drive sustainable development goals? *Technological Forecasting and Social Change, 199*(123082).
Makpotche, M., Bouslah, K., & M'Zali, B. (2024). Corporate governance and green innovation: International evidence. *Review of Accounting and Finance, 23*(2), 280–309.
Mhlanga, D., & Dzingirai, M. (2024). Bibliometric study on organizational resilience: Trends and future research agenda. *International Journal of Corporate Social Responsibility, 9*(1), 9.
Mutanda, V., Chirodzero, T. C., & Dzingirai, M. (2024). The role of green leadership in promoting responsible production. In *Responsible business and sustainable development* (pp. 200–218). Routledge.
Pamungkas, I. D., Raihan, M. R., Satata, D. P. I., & Kristianto, A. Y. (2024). Impact of corporate social responsibility between green accounting and sustainable development goals. *Jurnal Dinamika Akuntansi, 16*(1).
Pasha, A. T., Aiwan, A., & Akhtar, T. (2024). The impact of green governance on sustainability reporting & corporate social responsibility: Financial risk as moderator. *Journal of Social Research Development, 5*(2), 119–129.
Pathiranage, H. S. K., Anwer, S., Altaf, M., Ali, M. F., & Raza, M. F. (2024). Unlocking green leadership for effective performance and SDGs of organisation and project. *Educational Administration: Theory and Practice, 30*(6), 3931–3937.
Patnaik, H. K. (2024). Role of corporate good governance in achieving environmental sustainability towards fulfilling SDG agenda-2030 1.
Pham, N., Zhou, H., & Man, D. C. (2021). Corporate governance and corporate social responsibility: The role of executive compensation structure. *Journal of Business Ethics, 173*(1), 109–129.
Phan, T. C. (2024). Impact of green investments, green economic growth and renewable energy consumption on environmental, social, and governance practices to achieve the sustainable development goals: A sectoral analysis in the ASEAN economies. *International Journal of Engineering Business Management, 16*, 18479790241231725.

Pies, I., & Valentinov, V. (2024). Trade-offs in stakeholder theory: An ordonomic perspective. *Social Responsibility Journal, 20*(5), 975–997.

Purbawati, H. N. U., & Mujiyati, M. (2024). Pengaruh green accounting, corporate social responsibility dan good corporate governance Terhadap Profitabilitas Perusahaan Manufaktur Yang Terdaftar di BEI (Tahun 2018–2021). *Management Studies and Entrepreneurship Journal (MSEJ), 5*(2), 5300–5310.

Puspitasari, R. I., Saifi, M., & Nuzula, N. F. (2024). Competitive advantage as a mediation of the influence of green banking and corporate governance on company value. *Profit: Jurnal Adminsitrasi Bisnis, 18*(2), 168–184.

Saha, S., Hasan, A. R., Islam, K. R., & Priom, M. A. I. (2024). Sustainable Development Goals (SDGs) practices and firms' financial performance: Moderating role of country governance. *Green Finance, 6*(1), 162–198.

Salihi, A. A., Ibrahim, H., & Baharudin, D. M. (2024). Environmental governance as a driver of green innovation capacity and firm value creation. *Innovation and Green Development, 3*(2), 100110.

Sanusi, A., & Assih, P. (2024). The role of green accounting, good corporate governance, and environmental performance corporate social responsibility. *Indonesian Journal of Business Analytics, 4*(2), 643–652.

Sharma, E., & McLean, G. N. (2024). Corporate social responsibility, the United Nations' Sustainable Development Goals and financial performance, with implications for human resource development. *European Journal of Training and Development.*

Shin, S. J., Lee, J. H., & Kim, Y. K. (2022). Corporate governance, CSR reporting, and firm value: The mediating role of information asymmetry. *Journal of Business Ethics, 179*(1), 113–135.

Sitorus, R., & Sinaga, J. (2024). Regulations, policies, and governance in achieve sustainable development goals. *Indonesian Journal of Interdisciplinary Research in Science and Technology, 2*(5), 893–906.

Sungkawati, E. (2024). Opportunities and challenges: Adopting "Blue-Green Economy" terms to achieve SDGs. *Revenue Journal: Management and Entrepreneurship, 2*(1), 01–13.

Usman, B. M., Johl, S. K., & Khan, P. A. (2024). Fusion of green governance for sustainable development and world ecology: A tempting systematic review and bibliometric analysis. *Journal of Open Innovation: Technology, Market, and Complexity, 10*, 100309.

Valentinov, V., & Iliopoulos, C. (2024). The idea of adaptation in transaction cost economics: An application to stakeholder theory. *Society and Business Review* (ahead-of-print).

Wang, F., & Feng, H. (2024). Does incentive-based green governance compensate for green innovation in enterprises? The role of green orientation. *Environmental Science and Pollution Research, 31*(5), 7443–7464.

Wang, L., Ur Rehman, A., Xu, Z., Amjad, F., & Ur Rehman, S. (2023). Green corporate governance, green finance, and sustainable performance nexus in Chinese SMEs: A mediation moderation model. *Sustainability, 15*(13), 9914.

Yang, Z., Na, J., & Dong, X. (2024). Corporate governance for sustainable development: A study on mechanism configuration. *Journal of Cleaner Production, 458*, 142509.

Yanto, H., Pujiati, A., Handayani, B. D., Ridzuan, A. R., Keshminder, J. S., & Rachman, M. A. (2024). Good government governance as a moderator in achieving sustainable development goals in Indonesia. *International Journal of Sustainable Development & Planning, 19*(2).

Yu, W., & Choi, J. S. (2023). The role of institutional investors in green corporate governance: Evidence from CEO compensation contracts. *Journal of Corporate Finance, 73*, 101976.

Zhang, Y., Cao, S., Lin, X., Su, Z., & Wang, K. (2024). Corporate financial decision under green credit guidelines: Evidence from China. *Journal of the Asia Pacific Economy, 29*(2), 882–907.

CHAPTER 4

Evaluating Green Marketing Strategies: Role of Sustainability Indicators

Siphiwe Plaxcedes Mandina, Shelton Giwa, and Maximillian Chikwanha

Introduction

In today's business setting, sustainability and environmental stewardship have become more than just buzzwords; they are integral to corporate strategy and consumer preference (World Economic Forum, 2022). With increasing awareness of climate change and environmental degradation, both consumers and businesses are placing a greater emphasis on sustainability. Firms across various industries are adopting green marketing strategies to promote eco-friendly products, reduce carbon footprints, and enhance corporate social responsibility (CSR) (Ara et al., 2020). This shift is not only a response to regulatory pressures and market trends but also a

S. P. Mandina (✉) • S. Giwa • M. Chikwanha
Faculty of Business and Law, Leicester Castle Business School,
De Montfort University, Leicester, UK
e-mail: siphiwe.mandina@dmu.ac.uk; shelton.giwa2@dmu.ac.uk;
max.chikwanha@dmu.ac.uk

A. Muposhi, B. Nyagadza (eds.), *Sustainable Green Marketing Strategies for a Circular Economy*, Palgrave Studies of Marketing in Emerging Economies,
https://doi.org/10.1007/978-3-031-81627-7_4

reflection of a deeper commitment to preserving the planet for future generations.

Green marketing is a combination of marketing strategies that involve promoting products or services based on their environmental benefits, that is, consumer environmental and sustainability concerns (Paiva, 2020). This can range from highlighting the use of sustainable materials and processes to showcasing efforts to reduce emissions and waste. However, the effectiveness of these green initiatives is often questioned (Delmas & Pekovic, 2018). Despite the good intentions behind the application of green marketing, there is a growing concern about greenwashing, which is a practice where firms exaggerate or falsify their environmental claims to appear more sustainable than they are. Greenwashing can lead to consumer mistrust, damaging a company's reputation and undermining the efforts of genuinely sustainable businesses.

The need for a reliable framework to assess the impact and effectiveness of green marketing strategies is more pressing than ever. This chapter aims to address this need by evaluating green marketing strategies through the lens of sustainability indicators. Sustainability indicators are measurable metrics that provide concrete evidence of a company's environmental performance. These indicators can range from carbon footprint reduction and energy efficiency to waste management and water conservation. By focusing on these indicators, we can move beyond superficial claims and assess the true effectiveness of green marketing efforts.

The contemporary business landscape is increasingly defined by a growing emphasis on sustainability and environmental stewardship. This shift towards sustainability is driven by multiple factors. Consumers are becoming more environmentally conscious and demanding more sustainable products. Governments and regulatory bodies are implementing stricter environmental regulations. Investors are also recognizing the long-term value of sustainable business practices and incorporating environmental, social, and governance (ESG) criteria into their investment decisions. In this context, green marketing has emerged as a key strategy for businesses to differentiate themselves and gain a competitive edge.

However, the journey towards sustainable business practices is fraught with challenges. One of the main challenges is the lack of standardization in sustainability indicators. Different industries and firms use different metrics to measure their environmental performance, making it difficult to compare and evaluate their green marketing efforts. Moreover, the complexity of global supply chains adds another layer of difficulty in tracking and managing sustainability indicators. Despite these challenges, some businesses have made significant strides in integrating sustainability into their core operations and marketing strategies.

The theoretical foundation for evaluating green marketing strategies is based on several key concepts: sustainability, CSR, and stakeholder theory. Sustainability encompasses environmental, social, and economic dimensions, striving for a balance that ensures long-term viability. CSR integrates ethical considerations into business practices, encouraging firms to take responsibility for their impact on society and the environment. Stakeholder theory suggests that businesses should consider the interests of all stakeholders, including customers, employees, suppliers, communities, and the environment. This theoretical framework supports the analysis of green marketing strategies and the role of sustainability indicators in measuring their success.

Wide-ranging literature exists on green marketing and sustainability, providing valuable insights into the principles, tools, and challenges of sustainable branding. Ottman (2017) highlights the principles and tools for sustainable branding, emphasizing the importance of aligning marketing strategies with broader sustainability goals. Belz and Peattie (2021) provide a global perspective on sustainable marketing, offering practical guidelines for integrating sustainability into marketing practices. Ginsberg and Bloom (2004) examine strategic choices in green marketing, highlighting the need for businesses to understand consumer behaviour and preferences. In their study, White and Hardisty (2019) review strategies to shift consumer behaviour towards sustainability, underscoring the importance of effective communication and engagement.

Nevertheless, despite the existing knowledge and insights from the literature, there is still a need for a systematic approach to evaluating green marketing strategies. Many studies focus on individual aspects of green marketing or specific industries, but there is a lack of comprehensive frameworks that can be applied across different sectors. This chapter underscores the critical role of sustainability indicators in evaluating green marketing strategies. By adopting a systematic approach to measurement, businesses can enhance their environmental performance, build consumer trust, and achieve sustainable growth. The findings provide valuable insights for both academics and practitioners aiming to promote sustainability through effective marketing practices. As the business landscape continues to evolve, the importance of sustainability and green marketing will only grow, making it essential for businesses to adopt a comprehensive and transparent approach to their sustainability efforts.

This chapter seeks to achieve the following objectives:

- To understand the relationship between business performance, consumer behaviour, and environmental conservation.
- To identify some key indicators that can be applied in measuring sustainability within a company's marketing portfolio.
- To understand the importance of sustainability indicators for organizations in pursuit of green marketing efforts.

Theoretical Framework Development

Green marketing is more than just selling products; it's about promoting a way of doing business that respects the planet and its resources. It involves developing and promoting products and services that are not only high quality but also environmentally sustainable (Paiva, 2020). This means that these products and services are designed and produced in a way that minimizes negative impact on the environment. Firms that engage in green marketing are committed to reducing their carbon footprints, conserving natural resources, and fostering a more sustainable future (Urhan et al., 2023).

The theoretical foundation for evaluating green marketing strategies rests on several key concepts: sustainability, CSR, and stakeholder theory (Awa et al., 2024). These concepts are essential for understanding how businesses can effectively implement and measure their green marketing efforts. Sustainability is at the heart of green marketing. It encompasses three main dimensions: environmental, social, and economic (Correia, 2019; Purvis et al., 2019). The environmental dimension focuses on the impact that business activities have on the natural world, including issues like pollution, resource depletion, and biodiversity loss (Purvis et al., 2019). The social dimension considers how business practices affect people, including employees, customers, and communities (Kiyak & Grigoliene, 2023). This can involve aspects such as fair labour practices, community engagement, and consumer safety. The economic dimension looks at the financial performance of a business and how it can achieve long-term viability while still adhering to sustainable practices (Montiel-Hernández et al., 2024). True sustainability requires a balance between these three dimensions, ensuring that economic growth does not come at the expense of the environment or society (Dyllick & Hockerts, 2002).

CSR is another critical component of green marketing. CSR involves integrating ethical considerations into business practices. This means that firms take responsibility for the social and environmental impacts of their operations. They go beyond simply complying with laws and regulations to proactively improve their communities and the environment. CSR initiatives can include everything from reducing greenhouse gas emissions and conserving water to improving working conditions and supporting local charities. By incorporating CSR into their operations, firms can build trust with consumers and create a positive brand image that reflects their commitment to doing good (Carroll, 2019).

Stakeholder theory suggests that businesses should consider the interests of all stakeholders, not just shareholders. Stakeholders include anyone who is affected by a company's actions, such as customers, employees, suppliers, communities, and the environment. According to stakeholder theory, businesses have a responsibility to balance the needs and interests of these different groups. This approach recognizes that long-term business success depends on maintaining good relationships with all stakeholders, not just maximizing profits for shareholders. By considering the broader impact of their actions, businesses can create value for everyone involved and contribute to a more sustainable and equitable world (Freeman, 2014).

In this theoretical framework, sustainability, CSR, and stakeholder theory support the analysis of green marketing strategies and the role of sustainability indicators in measuring their success. Sustainability indicators are measurable metrics that provide concrete evidence of a company's environmental performance. They can include a wide range of data points, such as carbon footprint reduction, energy efficiency, waste management, water usage, and the use of renewable resources. By using these indicators, businesses can track their progress towards sustainability goals and make data-driven decisions to improve their practices (Global Reporting Initiative, 2013). For example, a company might measure its carbon footprint to understand how much greenhouse gas it is emitting. By tracking this indicator over time, the company can identify areas where it can reduce emissions, such as by improving energy efficiency or switching to renewable energy sources. Similarly, a company might measure its water usage to identify opportunities for conservation, such as by recycling water or using more efficient irrigation systems. By using sustainability indicators, businesses can demonstrate their commitment to environmental stewardship and provide transparent, verifiable evidence of their efforts (Epstein & Buhovac, 2014).

Green marketing involves developing and promoting products and services that are environmentally sustainable. The theoretical foundation for evaluating these strategies is based on sustainability, CSR, and stakeholder theory. These concepts provide a comprehensive framework for analysing green marketing efforts and understanding the role of sustainability indicators in measuring their success. By integrating these principles into their operations, businesses can enhance their environmental performance, build trust with consumers, and contribute to a more sustainable future.

Literature Development/Related Studies

The literature on green marketing and sustainability is vast and diverse, reflecting the growing importance of these topics in both academic research and practical business applications. This section reviews key studies and contributions that have shaped our understanding of green marketing, its principles, strategies, and the critical role of sustainability indicators.

Evolution of Green Marketing

A critical assessment accords the evolution of green marketing, tracing its development from the early days of environmental awareness to its current status as a mainstream business practice (Peattie & Crane, 2005). They highlight the shift from simple green product offerings to more sophisticated strategies that integrate sustainability into the entire business model. They also address the issue of greenwashing, where firms make exaggerated or false claims about the environmental benefits of their products. They emphasize the need for transparency and accountability in green marketing, arguing that long-term success depends on building genuine trust with consumers.

Global Perspective on Sustainability Marketing

The research provides a global perspective on sustainability marketing, highlighting the differences and similarities in how various regions approach green marketing. Their work underscores the importance of cultural context in shaping consumer attitudes towards sustainability. Belz and Peattie (2021) explore how sustainability marketing can be adapted to different markets, stressing the need for localization in messaging and strategy. Their book also delves into the challenges and opportunities businesses face when implementing green marketing strategies globally, offering practical advice on how to overcome barriers and leverage local strengths.

Principles and Tools for Sustainable Branding

Ottman (2017) offers a comprehensive discussion on the principles and tools essential for sustainable branding. Ottman's work emphasizes the need for authenticity in green marketing claims and the integration of sustainability into the core values of a brand. Ottman outlines strategies for creating sustainable products that appeal to environmentally conscious consumers, stressing that successful green branding goes beyond superficial changes. Ottman (2017) argues for a holistic approach where every aspect of the product lifecycle, from sourcing materials to disposal, is managed with sustainability in mind. Ottman's principles serve as a foundational guide for businesses seeking to develop credible and effective green marketing strategies.

Strategic Choices in Green Marketing

The strategic choices businesses must make when developing green marketing campaigns are examined by Ginsberg and Bloom (2004). The authors discuss the spectrum of green marketing strategies, from those that involve minor modifications to existing products, to those that require fundamental changes in business operations. Ginsberg and Bloom (2004) highlight the importance of aligning green marketing strategies with overall business objectives and the competitive landscape. They also argue that businesses need to carefully consider the trade-offs involved in pursuing green initiatives, such as potential higher costs and the need for consumer education.

Shifting Consumer Behaviour

White and Hardisty (2019) focus on strategies to shift consumer behaviour towards sustainability. Their research reviews various approaches, including nudging, social norms, and incentives, that can encourage consumers to make more sustainable choices. White and Hardisty (2019) stress the importance of understanding the psychological factors that drive consumer behaviour, such as the desire for social approval and the perception of personal efficacy. They argue that effective green marketing must address these underlying motivations and provide consumers with clear, actionable information on how they can contribute to sustainability.

Importance of Measurable Sustainability Indicators

A common theme on the importance of measurable sustainability indicators in evaluating the effectiveness of green marketing strategies has emerged across some studies. Scholars (e.g. Belz & Peattie, 2021; Ottman, 2017) emphasize the need for clear metrics to track progress and demonstrate the impact of green initiatives. Furthermore, scholars (Ginsberg & Bloom, 2004; Peattie & Crane, 2005) examine how sustainability indicators (such as carbon footprint reduction, energy efficiency, waste management, and water usage) can help businesses avoid greenwashing by providing verifiable evidence of their environmental claims. White and Hardisty (2019) further discuss how these indicators can be used to inform and motivate consumers, making sustainability efforts more tangible and credible. These contributions collectively underscore the necessity of integrating sustainability indicators into green marketing strategies to address consumer scepticism and enhance the credibility of environmental claims. By doing so, businesses can build trust with consumers.

Sustainability Indicators

Owing to the increasing complexity in the relationship between organizations and the environment, there has been a call for a conscious conservation of the environment through measurable operations (Walecka-Jankowska & Mrzygłocka-Chojnacka, 2023). As a result, organizations are becoming innovative in their approach to building environmentally friendly systems and sustainable business development models. Notably, organizations with high green indicators are more inclined towards the adoption of green marketing strategies (Esangbedo et al., 2024). Manufacturing firms, for example, should be more poised towards eco-innovation in order to garner effective sustainability outcomes (Esangbedo et al., 2024). Global sustainability indicators such as the Global Reporting Initiative (GRI) are the epitome of worldwide best practices on economic, environmental, and social impacts (Narula et al., 2021). This is because such indicators as the GRI have facilitated the development of plausible solutions to deal with climate change in the long term, especially for emerging economies (Mougenot & Doussoulin, 2024). Worth noting is that, to a greater extent, Industry 4.0 influences the adoption of economic, environmental, and social GRI standards. This is therefore a call to action for researchers and practitioners to be more comprehensive and rigorous in testing and validating the GRI as a significant and effective sustainability indicator. In support of this, the GRI framework is widely acknowledged and used by Sri Lankan organizations that have mostly relied on its efficacy and flexibility in continuous improvement (Dissanayake, 2021). Thus, in accordance with the 2030 Sustainable Development Goals, businesses need to adapt to the principles of a green economy (Chen et al., 2022).

Research Method

In order to explore the effectiveness of green marketing strategies through the lens of sustainability indicators, a thematic literature review was assessed using the Preferred Reporting Items for Systematic reviews and Meta-Analyses (PRISMA) Statement. Existing literature was synthesized, key themes were identified, and meaningful insights were drawn from a wide array of studies. This approach allows for a comprehensive understanding of the current state of knowledge and highlights areas where further research is needed.

Research Context and Data Collection

The context for this research spans multiple industries that have adopted green marketing strategies, including consumer goods, automotive, energy, technology, and more. These sectors share a common commitment to reducing environmental impact and promoting sustainable practices through their marketing efforts. This diverse backdrop is crucial for evaluating the effectiveness of green marketing strategies across different contexts and industries (Belz & Peattie, 2021). The target population for this review includes firms that have implemented green marketing strategies and the consumers who interact with these firms. This dual focus ensures that the analysis captures both the corporate perspective and the consumer response, which is critical for assessing the overall effectiveness of green marketing initiatives (Ginsberg & Bloom, 2004).

The sample for this thematic literature review consists of peer-reviewed journal articles, books, industry reports, and case studies focusing on green marketing and sustainability indicators. The selection criteria for these sources include relevance to the research questions, publication within the last two decades to ensure currency, and credibility of the sources, such as publications in reputable journals or by well-known authors in the field (Ottman, 2017; Peattie & Crane, 2005). To ensure analytical rigour, this thematic literature review employs a systematic approach to identifying, evaluating, and synthesizing relevant literature. The process begins with a comprehensive search of academic databases such as Google Scholar, JSTOR, and ScienceDirect using keywords like green marketing, sustainability indicators, CSR, and stakeholder theory (White & Hardisty, 2019). Once a broad range of sources is gathered, each source is evaluated for its relevance and contribution to the research questions.

Data Analysis

The data analysis process involves several steps. First, each selected source is read thoroughly to identify key themes and insights related to green marketing and sustainability indicators. These themes are then categorized into broader groups, such as the principles of green marketing, strategic implementation, consumer behaviour, and the measurement of sustainability impacts (Belz & Peattie, 2021). Thematic coding is used to organize the data, allowing for the identification of patterns and trends across different studies. This process helps to highlight commonalities and differences in how various researchers approach the topic, as well as gaps in the existing literature. For instance, themes such as the challenges of greenwashing, the importance of transparency, and the role of consumer education frequently emerge (Peattie & Crane, 2005).

By synthesizing these themes, the review provides a nuanced understanding of the current landscape of green marketing. It identifies best practices, common pitfalls, and the critical role that measurable sustainability indicators play in ensuring the authenticity and effectiveness of green marketing efforts (Ottman, 2017). The thematic literature review method used in this research provides a robust framework for evaluating green marketing strategies. By systematically identifying and analysing key themes from a diverse range of sources, this approach offers a comprehensive understanding of the effectiveness of green marketing initiatives. The thematic literature review also highlights the importance of sustainability indicators in measuring success and addresses critical issues such as greenwashing and consumer scepticism. Through this method, the research contributes valuable insights that can guide businesses, policymakers, and researchers in advancing sustainable marketing practices (White & Hardisty, 2019).

Results

The thematic literature review yielded several significant findings regarding the effectiveness of green marketing strategies and the role of sustainability indicators. These findings are based on an extensive analysis of various studies and sources, allowing for a comprehensive understanding of the current landscape of green marketing.

Key Themes Identified

The literature emphasizes the foundational principles of green marketing, which include the development and promotion of environmentally sustainable products and services. These principles are rooted in the broader concept of sustainability, encompassing environmental, social, and economic dimensions (Belz & Peattie, 2021). Firms that adhere to these principles are more likely to gain consumer trust and achieve long-term success. Effective green marketing requires strategic implementation, which involves integrating sustainability into the core business operations and marketing strategies. This includes adopting practices such as reducing carbon footprints, enhancing energy efficiency, and promoting the use of renewable resources.

The literature highlights that firms with well-defined sustainability goals and strategies are more successful in their green marketing efforts (Ottman, 2017). Understanding consumer behaviour is crucial for the success of green marketing strategies. The review shows that consumers are increasingly aware of and concerned about environmental issues. However, there is also a significant level of scepticism towards green claims due to instances of greenwashing, where firms falsely promote their products as environmentally friendly (Peattie & Crane, 2005). Therefore, transparency and authenticity are essential in green marketing to build consumer trust.

The role of sustainability indicators is a critical theme in the literature. These indicators provide measurable metrics that can track a company's environmental performance and the effectiveness of its green marketing strategies. Common sustainability indicators include carbon footprint reduction, energy efficiency, waste management, and water usage. The literature underscores the importance of using these indicators to demonstrate tangible progress towards sustainability goals (White & Hardisty, 2019).

Best Practices and Common Pitfalls

The review identifies several best practices in green marketing, such as:

Transparency in Reporting

Firms that are transparent about their sustainability efforts and provide verifiable data tend to gain more consumer trust and loyalty (Ottman, 2017). Effective green marketing strategies involve engaging various stakeholders, including employees, customers, suppliers, and the community. This engagement helps in fostering a collective commitment to sustainability (Ginsberg & Bloom, 2004). Firms that continuously evaluate and improve their sustainability practices are more likely to achieve long-term success. This involves regularly updating sustainability goals and strategies based on new insights and technological advancements (Peattie & Crane, 2005). Misleading consumers about the environmental benefits of a product or service can lead to significant reputational damage and loss of consumer trust (Peattie & Crane, 2005). Consumers may not fully understand the benefits of green products, leading to lower adoption rates. Therefore, educating consumers about the environmental impact of their choices is essential (White & Hardisty, 2019).

Importance of Sustainability Indicators

The findings highlight that sustainability indicators are crucial for evaluating the success of green marketing strategies. These indicators provide concrete evidence of a company's environmental performance and help in making data-driven decisions. For example, tracking carbon emissions can reveal areas where a company can improve its energy efficiency and reduce its environmental impact. Similarly, monitoring waste management practices can identify opportunities for reducing waste and promoting recycling (Ottman, 2017). The results of this thematic literature review provide a comprehensive understanding of the effectiveness of green marketing strategies. The identified themes and best practices highlight the importance of transparency, stakeholder engagement, and continuous improvement. Furthermore, the critical role of sustainability indicators in measuring the success of green marketing efforts is emphasized. These insights can guide businesses, policymakers, and researchers in advancing sustainable marketing practices and addressing the challenges associated with greenwashing and consumer scepticism.

The findings from this study underscore the pivotal role that sustainability indicators play in evaluating the effectiveness of green marketing strategies. Businesses that successfully integrate these indicators into their core operations not only improve their environmental performance but also gain a significant competitive advantage. However, the challenge of standardizing these indicators across different industries persists and is critical for ensuring consistent evaluation and comparison.

Enhancing Environmental Performance

One of the most notable benefits of integrating sustainability indicators into green marketing strategies is the enhancement of environmental performance. Firms that measure and monitor indicators such as carbon footprint, energy efficiency, and waste management can make informed decisions that lead to substantial environmental improvements (Lim, 2016). For instance, a company that tracks its carbon emissions can identify key areas where it can reduce energy consumption, switch to renewable energy sources, and thus lower its overall carbon footprint. Similarly, monitoring waste management practices can help a company implement more effective recycling and waste-reduction strategies.

Competitive Advantage

Beyond environmental benefits, the use of sustainability indicators also provides firms with a competitive edge. Consumers today are increasingly conscious of environmental issues and prefer to support brands that demonstrate a genuine commitment to sustainability. By transparently reporting on sustainability indicators, firms can build trust with consumers and differentiate themselves from competitors who may not be as forthcoming or genuine in their environmental claims. This transparency not only attracts environmentally conscious consumers but also enhances brand loyalty and reputation.

Challenges of Standardization

Despite the clear advantages, the challenge of standardizing sustainability indicators across different industries remains significant. Each industry has its unique characteristics and environmental impacts, making it difficult to develop a one-size-fits-all set of indicators. For example, the sustainability indicators relevant to the automotive industry, such as emissions from manufacturing and vehicle use, may differ considerably from those pertinent to the consumer goods industry, which might focus more on packaging waste and supply chain sustainability.

To address this challenge, there is a need for industry-specific guidelines that provide tailored sustainability indicators while still aligning with broader, universally accepted principles of sustainability. Standardizing these indicators would facilitate more accurate and consistent evaluations and comparisons across different sectors. It would also help in benchmarking performance, enabling firms to understand how they stack up against their peers and identify best practices that can be adopted or adapted to their specific contexts.

Importance of Transparency and Authenticity

The issue of greenwashing, where firms falsely claim or exaggerate their environmental efforts, emerges as a significant concern in the literature. The use of verifiable and transparent sustainability indicators can mitigate this risk by providing concrete evidence of a company's environmental performance. Transparency in reporting these indicators not only builds consumer trust but also holds firms accountable for their sustainability claims. This accountability is crucial for maintaining the integrity of green marketing efforts and ensuring that they lead to genuine environmental benefits.

Role of Stakeholder Engagement

Engaging stakeholders, including employees, customers, suppliers, and the broader community, is another critical aspect highlighted in the findings. Effective green marketing strategies are those that involve stakeholders in the sustainability journey. This engagement can take various forms, such as involving employees in sustainability initiatives, educating consumers about the environmental impact of their choices, and collaborating with suppliers to enhance sustainability throughout the supply chain (Ng & Wang, 2020). By fostering a culture of sustainability within and outside the organization, firms can ensure that their green marketing efforts are holistic and impactful.

The discussion reaffirms the essential role of sustainability indicators in evaluating and enhancing green marketing strategies. While the challenges of standardization across industries are significant, addressing these issues is crucial for ensuring consistent and meaningful evaluations. By integrating sustainability indicators into their operations, maintaining transparency, and engaging stakeholders, businesses can not only improve their environmental performance but also gain a competitive advantage in the increasingly sustainability-conscious market.

Implications

The findings of this study have several important implications for theory, policy, and practice. They highlight the critical role of measurable sustainability indicators in the realm of green marketing and provide valuable insights for businesses, policymakers, and researchers.

Theoretical Implications

From a theoretical perspective, this research extends the understanding of green marketing by emphasizing the necessity of measurable sustainability indicators. Previous theories on green marketing have focused broadly on promoting environmentally friendly practices and products. However, this study underscores the importance of having concrete, quantifiable metrics to evaluate the success and authenticity of these strategies. By integrating sustainability indicators into theoretical frameworks, researchers can develop more robust models that accurately assess the impact of green marketing efforts. This refined understanding helps in distinguishing genuine sustainability initiatives from superficial greenwashing practices, thereby advancing the field of green marketing research.

Policy Implications

The policy implications of this study are equally significant. Policymakers play a crucial role in shaping the regulatory environment in which businesses operate. This research suggests that policymakers should promote the development and adoption of standardized sustainability indicators and reporting frameworks. Standardization would ensure that all firms are held to the same criteria, making it easier to compare their environmental performance. This consistency would enhance transparency and accountability, helping to prevent greenwashing and ensuring that firms are genuinely contributing to environmental sustainability. Moreover, clear guidelines and standards would provide businesses with a roadmap for implementing effective green marketing strategies, ultimately benefiting the broader goal of sustainable development.

For businesses, the practical implications of this research are profound. Firms can use the insights gained from this study to refine their green marketing strategies, making sure they are not only environmentally sustainable but also effectively communicated to stakeholders. By adopting measurable sustainability indicators, businesses can track their progress more accurately and make data-driven decisions to improve their practices (Ginsberg & Bloom, 2004). This transparency in reporting sustainability efforts can build trust with consumers, investors, and other stakeholders, enhancing the company's reputation and competitive position.

In addition, businesses can use these indicators to identify areas where they can make the most significant environmental impact. For example, a company might find that improving energy efficiency in its operations leads to substantial reductions in carbon emissions, which can then be highlighted in its marketing campaigns. By focusing on genuine improvements and transparently reporting these efforts, businesses can differentiate themselves from competitors and build a loyal customer base that values sustainability.

Another practical implication is the potential for increased collaboration between businesses and other stakeholders. By engaging with consumers, employees, suppliers, and community groups, firms can foster a culture of sustainability that goes beyond marketing. This collaborative approach can lead to innovative solutions and shared best practices, further enhancing the effectiveness of green marketing strategies. For instance, involving suppliers in sustainability initiatives can lead to more environmentally friendly supply chains, while educating consumers about sustainable choices can drive demand for greener products.

The implications of this research are wide-ranging and impactful. Theoretically, it enriches the understanding of green marketing by highlighting the importance of measurable sustainability indicators. For policymakers, it underscores the need for standardized frameworks to ensure transparency and accountability. Practically, it provides businesses with actionable insights to refine their green marketing strategies and build stronger, trust-based relationships with their stakeholders. By adopting these recommendations, businesses and policymakers can contribute to a more sustainable future, where green marketing efforts are authentic, effective, and widely recognized.

Study Limitations

While this study offers valuable insights into the role of sustainability indicators in evaluating green marketing strategies, it is essential to acknowledge its limitations. Understanding these limitations helps contextualize the findings and suggests avenues for future research.

One of the primary limitations of this study is its focus on a relatively narrow selection of firms and industries. Although the research covers various sectors such as consumer goods, automotive, energy, and technology, it does not encompass the full spectrum of industries that might engage in green marketing. Each industry has its unique characteristics and environmental challenges, which means the findings may not be universally applicable. For instance, the strategies effective in the consumer goods sector may differ significantly from those in heavy manufacturing or agriculture. A broader range of industries would provide a more comprehensive picture of green marketing practices and their effectiveness.

Scope of Literature Review

While the thematic literature review method allows for an in-depth analysis of existing research, it inherently limits the number of sources and cases examined (Delmas & Pekovic, 2018). A larger sample size, including more firms and a wider variety of case studies, would strengthen the reliability and generalizability of the findings. Future studies should aim to include a more extensive dataset, possibly incorporating quantitative surveys and empirical data from a larger pool of firms to validate the insights gained from the literature review.

Geographical Constraints

The geographical scope of the reviewed literature also poses a limitation. The majority of the sources may focus on green marketing practices in developed countries, where sustainability initiatives are often more advanced and better documented. This focus might overlook the practices and challenges faced by firms in developing regions, where environmental awareness and resources for sustainable practices might differ significantly. Including a more diverse geographical representation in future research would provide a more global perspective on green marketing and sustainability indicators.

Variability in Sustainability Indicators

Another limitation lies in the variability of sustainability indicators used across different studies and industries. There is no universal set of indicators, and the metrics can vary widely depending on the sector, company size, and regional regulations. This variability makes it challenging to compare and standardize findings across different contexts. Future research may consider working towards developing a more consistent set of sustainability indicators that can be applied universally or at least adapted to specific industry standards, facilitating more straightforward comparisons and assessments.

Potential Biases in Literature Selection

The thematic literature review method, while systematic, may still be subject to selection biases. The process of choosing relevant literature can be influenced by the availability of sources, the researchers' perspectives, and the inherent biases in the existing body of research. To mitigate this, future studies should strive for greater inclusivity and transparency in literature selection, possibly incorporating a broader range of sources, including grey literature and unpublished studies.

Dynamic Nature of Green Marketing

Finally, green marketing is a dynamic and rapidly evolving field. The strategies and sustainability indicators that are effective today may change as new technologies emerge, consumer preferences shift, and regulations evolve. This study provides a snapshot based on current literature, but continuous research is necessary to keep up with these changes. Longitudinal studies that track the evolution of green marketing practices over time would be particularly valuable in capturing these dynamics.

While this study provides significant insights into the role of sustainability indicators in green marketing, its limitations highlight the need for further research. Expanding the scope to include more industries, a larger sample size, diverse geographical contexts, and standardized sustainability indicators would enhance the comprehensiveness and applicability of future findings. By addressing these limitations, future research can build on this foundation to offer deeper, more actionable insights into the effectiveness of green marketing strategies.

Conclusion

In today's business environment, where sustainability has become more than just a buzzword, the role of sustainability indicators in evaluating green marketing strategies cannot be overstated. This chapter has thoroughly examined how these indicators serve as essential tools for businesses to not only gauge their environmental impact but also to enhance their performance, build trust with consumers, and drive sustainable growth. By taking a systematic approach to measurement, firms can turn their green marketing efforts from mere rhetoric into tangible, impactful actions.

Sustainability indicators are metrics that provide concrete evidence of a company's environmental, social, and economic performance (Carroll, 2019). They help businesses track progress towards their sustainability goals and make informed decisions. Indicators such as carbon footprint reduction, energy efficiency, water usage, waste management, and the use of renewable resources are critical for understanding and improving a company's environmental impact. Without these indicators, it would be challenging to measure the effectiveness of green marketing strategies or to demonstrate a company's commitment to sustainability.

One of the primary benefits of using sustainability indicators is the ability to enhance environmental performance. By systematically measuring and tracking various aspects of their operations, firms can identify areas where they can reduce their environmental footprint. For instance, a company might discover through its indicators that it is using more energy than necessary in its manufacturing processes. By addressing this issue, the company can improve its energy efficiency, lower its carbon emissions, and reduce costs. This not only benefits the environment but also contributes to the company's bottom line (Epstein & Buhovac, 2014).

In the era of conscious consumerism, where customers are increasingly aware of and concerned about environmental issues, building trust is paramount. Consumers today are not easily swayed by marketing claims; they demand transparency and proof of a company's environmental efforts. Sustainability indicators provide this proof. By publicly sharing their sustainability metrics and progress, firms can demonstrate their commitment to the environment and earn the trust of their consumers. This transparency helps to counteract scepticism and accusations of greenwashing, where firms are perceived to exaggerate or falsely claim their environmental efforts.

Sustainable growth is about achieving business success without compromising the ability of future generations to meet their needs. This balance requires a long-term perspective and a commitment to continuous improvement in sustainability practices. Sustainability indicators play a crucial role in this by providing a clear picture of where a company stands and where it needs to go. Businesses that effectively integrate these indicators into their core operations are better positioned to achieve sustainable growth. They can align their strategies with their sustainability goals, create value for all stakeholders, and ensure their longevity in an increasingly eco-conscious market (Freeman, 2014).

The findings of this chapter offer valuable insights for both academics and practitioners. For academics, the detailed examination of sustainability indicators in the context of green marketing provides a foundation for further research. There is still much to explore regarding the standardization of these indicators, their application across different industries, and their impact on various aspects of business performance. Future research could delve deeper into these areas, contributing to a more comprehensive understanding of how sustainability indicators can drive green marketing success.

For practitioners, the insights gained from this study can guide the development and implementation of more effective green marketing strategies. By understanding the importance of sustainability indicators, businesses can refine their approaches to ensure they are genuinely sustainable and resonate with consumers. Practical steps include setting clear sustainability goals, choosing the right indicators to measure progress, and being transparent about their efforts and achievements. This approach not only helps to build consumer trust but also fosters a culture of accountability and continuous improvement within the organization.

One of the significant challenges in green marketing is overcoming greenwashing and consumer scepticism. Greenwashing occurs when firms mislead consumers about the environmental benefits of their products or services. This can damage a company's reputation and erode consumer trust. Sustainability indicators help to mitigate this risk by providing verifiable data that backs up a company's claims. When firms use standardized and widely recognized indicators, they can more effectively communicate their genuine commitment to sustainability, thereby reducing scepticism and building a more loyal customer base.

A critical takeaway from this chapter is the need for standardization of sustainability indicators across industries. Currently, the lack of universal standards makes it difficult to compare and evaluate green marketing efforts consistently. Developing and adopting standardized indicators would allow for more reliable benchmarking and assessment. This would not only benefit businesses in measuring and improving their sustainability performance but also help consumers make more informed decisions based on comparable data. Policymakers and industry groups have a crucial role to play in driving this standardization, promoting transparency, and ensuring that sustainability claims are credible and meaningful.

While this chapter provides a comprehensive overview of the role of sustainability indicators in green marketing, it also highlights areas where further research and action are needed. Future studies should expand the scope to include more industries, a larger sample size, and diverse geographical contexts. This broader approach would provide a more comprehensive understanding of green marketing practices and their effectiveness. Additionally, research should continue to explore the development of standardized sustainability indicators and their application in different business contexts.

In practice, businesses should continue to innovate and improve their sustainability practices. This includes investing in new technologies, adopting best practices from other industries, and continuously engaging with stakeholders to understand and address their concerns. By doing so, firms can stay ahead of regulatory requirements, meet consumer expectations, and contribute to a more sustainable future.

References

Ara, H., Yeap, J. A. L., & Hassan, S. H. (2020). Adoption of green marketing strategy: A conceptual view based on stakeholder theory. In Z. Ahmad (Ed.), *Progressing beyond and better: Leading businesses for a sustainable future, vol. 88. European proceedings of social and behavioural sciences* (pp. 241–249). European Publisher.

Awa, H. O., Etim, W., & Ogbonda, E. (2024). Stakeholders, stakeholder theory and Corporate Social Responsibility (CSR). *International Journal of Corporate Social Responsibility, 9*(1), 11.

Belz, F. M., & Peattie, K. (2021). *Sustainability marketing: A global perspective*. Wiley.

Carroll, A. B. (2019). Corporate social responsibility: Evolution of a definitional construct. *Business & Society, 3*(38), 268–295.

Chen, M., Chen, L., Cheng, J., & Yu, J. (2022). Identifying interlinkages between urbanization and Sustainable Development Goals. *Geography and Sustainability, 3*(4), 339–346.

Correia, M. S. (2019). Sustainability: An overview of the triple bottom line and sustainability implementation. *International Journal of Strategic Engineering (IJoSE), 2*(1), 29–38.

Delmas, M. A., & Pekovic, S. (2018). Organizational configurations for sustainability and profitability: *A qualitative comparative analysis. Business & Society*, 216–251.

Dissanayake, D. (2021). Sustainability key performance indicators and the global reporting initiative: Usage and challenges in a developing country context. *Meditari Accountancy Research, 29*(3), 543–567.

Dyllick, T., & Hockerts, K. (2002). Beyond the business case for corporate sustainability. *Business Strategy and the Environment, 2*(11), 130–141.

Epstein, M. J., & Buhovac, A. R. (2014). *Making sustainability work: Best practices in managing and measuring corporate social, environmental and economic impacts*. Routledge.

Esangbedo, C.O., Chang, Y., Ushie, A. M., & Hasan, M. M. (2024). Eco-Innovation and Firm Sustainability in Emerging Economies: The Role of Consumer Behavior and Green B2B Marketing. *Journal of Business-to-Business Marketing, 31*(4), 405–420.

Freeman, R. E. (2014). *Strategic management: A stakeholder approach* (1st ed.). Cambridge University Press.

Ginsberg, J. M., & Bloom, P. N. (2004). Choosing the right green marketing strategy. *MIT Sloan Management Review, 1*(46), 79–84.

Global Reporting Initiative. (2013). *Sustainability reporting guidelines*. GRI.

Kiyak, D., & Grigoliene, R. (2023). Analysis of the conceptual frameworks of green marketing. *Sustainability, 15*(21), 15630.

Lim, W. M. (2016). A blueprint for sustainability marketing: Defining its conceptual boundaries for progress. *Marketing Theory*, 232–249.

Montiel-Hernández, M. G., Pérez-Hernández, C. C., & Salazar-Hernández, B. C. (2024). The intrinsic links of economic complexity with sustainability dimensions: A systematic review and agenda for future research. *Sustainability, 16*(1), 391.

Mougenot, B., & Doussoulin, J. P. (2024). A bibliometric analysis of the Global Reporting Initiative (GRI): global trends in developed and developing countries. *Environment, Development and Sustainability, 26*(3), 6543–6560.

Narula, S., Puppala, H., Kumar, A., Frederico, G.F., Dwivedy, M., Prakash, S., & Talwar, V. (2021). Applicability of industry 4.0 technologies in the adoption of global reporting initiative standards for achieving sustainability. *Journal of Cleaner Production, 305*, 127141.

Ng, A., & Wang, W. (2020). Understanding the role of sustainability in branding of consumer products. *Journal of Brand Management*, 55–72.

Ottman, J. A. (2017). *The new rules of green marketing: Strategies, tools, and inspiration for sustainable branding*. Berrett-Koehler Publisher.

Paiva, T. (2020). Green Marketing. In S. Idowu, R. Schmidpeter, N. Capaldi, L. Zu, M. Del Baldo, & R. Abreu (Eds.), *Encyclopedia of sustainable management*. Springer.

Peattie, K., & Crane, A. (2005). Green marketing: Legend, myth, farce or prophesy? *Qualitative Market Research: An International Journal, 8*(4), 357–370.

Purvis, B., Mao, Y., & Robinson, D. (2019). Three pillars of sustainability: In search of conceptual origins. *Sustainability Science, 14*, 681–695.

Urhan, B., Hoştut, S., Güdekli, İ. A., & Aydoğan, H. (2023). Climate change and marketing: A bibliometric analysis of research from 1992 to 2022. *Environmental Science and Pollution Research, 30*(34), 81550–81572.

Walecka-Jankowska, K., & Mrzygłocka-Chojnacka, J. (2023). Relationship between environmental innovation and sustainable outcomes–empirical evidence from Poland. Zeszyty Naukowe. Organizacja i Zarządzanie/ Politechnika Śląska.

White, K. H., & Hardisty, D. J. (2019). How to SHIFT consumer behaviors to be more sustainable: A literature review and guiding framework. *Journal of Marketing, 3*(83), 22–49.

World Economic Forum. (2022). Why sustainability is crucial for corporate strategy. Retrieved September 23, 2024, from https://www.weforum.org/agenda/2022/06/why-sustainability-is-crucial-for-corporate-strategy/#:~:text=Prioritizing%20sustainability%20in%20corporate%20strategy,acquisition%20and%20ensure%20increased%20productivity

CHAPTER 5

Underlying Drivers of Social Innovation Acceptance: A Case of Reusable Sanitary Towels

Asphat Muposhi , *Melissa Zulu* , *and Hazel Gwasira*

Introduction

Reusable sanitary towels are positioned as an environmentally friendly and affordable alternative to disposable sanitary towels (Sassone et al., 2022; Mehta et al., 2020). Poor menstrual health and hygiene is a cause of concern in developing countries as it has become a public health problem affecting many adolescent girls and women from disadvantaged backgrounds (Mantsebo, 2021). However, menstrual health and hygiene (MHH) needs are often overlooked, leaving adolescent girls and women

A. Muposhi (✉) • M. Zulu
Marketing Division, School of Business Sciences, University of the Witwatersrand, Johannesburg, South Africa
e-mail: asphat.muposhi@wits.ac.za; melissa.zulu@wits.ac.za

H. Gwasira
Harare, Zimbabwe
e-mail: hazel.gwasira@telone.ac.zw

A. Muposhi, B. Nyagadza (eds.), *Sustainable Green Marketing Strategies for a Circular Economy*, Palgrave Studies of Marketing in Emerging Economies,
https://doi.org/10.1007/978-3-031-81627-7_5

facing difficulties regarding access to adequate sanitary towels (Kambala et al., 2020). The dire economic situation in developing countries is making it hard for women from disadvantaged backgrounds to access basic sanitary wear (Mantsebo, 2021). Due to their inability to acquire sanitary towels, they often resort to makeshift and unclean menstrual products, which have been linked to urogenital infections (Kambala et al., 2020; Mauye, 2024). As a result, many girls and women from disadvantaged backgrounds experience anxiety, isolation, shame, and stigma as they approach their menstrual cycle (Tembo, 2022).

In Southern African countries such as Malawi, Zambia and Zimbabwe, the menstrual health of girls and women from disadvantaged backgrounds is worsened by limited facilities for water, sanitation and hygiene (Witte, 2021; Tembo, 2022). The use of unhygienic menstrual materials is associated with reproductive tract infections, rashes and cervical cancer (Bhebhe, 2021). Young adolescent girls have been reported as missing school every month because they could not afford disposable hygiene products. Moreover, the relatively high costs of commercial sanitary pads have to compete against other needed items in the family, such as food (Witte, 2021). Recently, reusable sanitary towels were introduced as a way of addressing the plight of girls and women from disadvantaged backgrounds (Sassone et al., 2022). However, there are mixed views related to the use of reusable sanitary towels (Bhebhe, 2021; Tembo, 2022). Against this background, this chapter examines rational and normative factors that influence the use of reusable sanitary towels.

This chapter seeks to address the following research objectives:

- To examine the influence of consumer attitudes towards intention to use reusable sanitary towels,
- To examine the relationship between subjective norms and intention to use reusable sanitary wear,
- To ascertain the relationship between perceived behavioural control and actual behaviour of using reusable sanitary wear, and
- To examine the relationship between intention and actual behaviour of using sanitary wear.

From a theoretical standpoint, this chapter intends to contribute to the body of knowledge by delineating the rational and normative drivers of social innovation acceptance in the case of reusable sanitary towels. This chapter also provides input to policymakers who intend to develop and implement comprehensive menstrual health management interventions. The next section reviews literature on reusable sanitary towels.

Literature Review

Acceptance of Reusable Sanitary Towels as Social Innovation

Reusable sanitary towels are positioned as an environmentally friendly and affordable alternative to disposable sanitary towels (Mehta et al., 2020). Reusable sanitary towels are available in the form of pads, pants and menstrual cups (Sassone et al., 2022). The Afripad from Uganda is one of the many new low-cost reusable sanitary towel innovations. The Afripad is marketed as a cloth-based product used during menstruation, produced locally from sewn materials that are created to be washed and reused and are sold directly to schoolgirls as well as distributed freely by non-governmental organisations (Tembo, 2022). According to Sassone et al. (2022), reusable sanitary towels are comfortable to wear, easy to clean and provide adequate absorption. Being as such, they allow adolescent girls to focus on school work and worry less about leaking and needing to find privacy to change menstrual materials.

Reusable sanitary towels are also considered comfortable and easy to use, as compared to their disposable counterparts, which are usually improperly disposed of (Mehta et al., 2020). Studies by Medina-Perucha (2022) revealed that women between the ages 18–55 in Spain preferred reusable sanitary towels, as they are perceived to be more acceptable than non-reusable ones. Barriers to use menstrual products were also identified as associated with experiences of menstrual inequality such as menstrual poverty and lack of access to information or menstrual management facilities (Medina-Perucha et al., 2022). Similarly, a report by the UNHCR (2019) revealed that the acceptance of reusable sanitary towels was driven by environmental consciousness and affordability.

More information on menstruation care needs to be included in puberty education by educators, and advocates should increase public understanding of the proper hygiene procedures for this product (Ramsay et al., 2023). Studies in six countries (India, Kenya, Uganda, Ethiopia and Zimbabwe) revealed that girls' and women's perceptions of what a menstruation material should give are altered by exposure to new goods, with receivers placing a high value on more mobility and less worry about leaking. Following the distribution of instructional materials or instruction on the utilisation of novel goods, females reported feeling more capable and knowledgeable about the physiological process involved in menstruation.

In certain instances, they expressed a desire to share this information with others in their communities (Shannon et al., 2021).

In a study conducted in Nigeria, Onasoga et al. (2022) reported poor perception and non-acceptance of reusable sanitary towels among secondary school adolescent girls. This attitude emanated from factors such as lack of awareness, stress of washing reusable menstrual towels, and risk of contracting infection. The findings correspond to those by van Eijk et al. (2021) who reported that the uptake of reusable sanitary towels was lower compared to cloth, pointing to reported concerns related to skin irritation, challenges with washing and changing the reusable towels. Singh et al. (2022) argue that reusable sanitary towels like the silicon cup are safe, eco-friendly, durable, and cheap; hence, their non-acceptability may be due to higher adoption barriers. As such, adaptability can increase gradually through proper counselling, peer support and practice.

Studies by Alexandra (2021) revealed that commercialisation of environmentally friendly menstrual products by social enterprises was a great solution to reduce the impact on the environment, but none of them were magic bullets to the MHH problems in India. These findings were also supported by Kumari and Kumar (2023) who analysed a case of Self Help Groups (SHGs) within the context of social entrepreneurship where a grassroots innovation was implemented for sustainable production of reusable sanitary towels and engaging in practices related to menstruation. The findings indicated that various challenges were faced by SHGs in implementing and running the reusable sanitary towel production unit due to social taboos, lack of infrastructure and other support systems from the formal sector in the region. In a study conducted in the context of Zimbabwe, Nyava-Machadu (2024) found that many girls endure menstrual poverty due to a lack of access to adequate menstrual hygiene products and facilities. Hence, this salient crisis not only affects the physical and mental well-being of women but also hampers their education and economic opportunities.

UNICEF (2019) reported that 94.7 percent of girls in urban and 85 percent of girls in rural areas of Zimbabwe commonly use disposable sanitary pads. However, girls struggle to manage their menstruation with dignity, as they do not have access to or cannot afford sanitary products due to high costs. According to Nyamayedenga and Tsvere (2020), the sustainable management of disposable sanitary towels continues to be a big problem for urban municipalities. Due to inadequate or nonexistent waste management systems, a large quantity of these absorbent hygiene products (AHPs) is being thrown in neighbourhoods. Tambo reports that approximately 76 percent of girls in both rural and urban schools lack sufficient understanding about menstruation and puberty. Consequently, they face shame and stigma, as well as low self-esteem because of their dread and anxiety during menstruation. Many of these women and girls do not have access to menstruation products that are both economical and effective, to secure locations where they can be cleaned or disposed of both within and outside the home, or to reliable and accurate information about menstrual health.

Conceptual Framework and Hypotheses Development

This chapter is underpinned by the theory of planned behaviour (Ajzen, 1991). Drawing from the theory of planned behaviour (TPB), this chapter argues that attitudes, subjective norms and perceived behavioural control influence intentions to use reusable sanitary towels. Additionally, intention is expected to influence the actual behaviour of using reusable towels.

Consumer Attitudes and Intention to Use Reusable Sanitary Towels

Attitudes refer to favourable or unfavourable feelings that influence behavioural intention (Ajzen, 1991). Intention describes a person's deliberate choice or goal to carry out a specific behaviour. Intention is construed as a proxy of willingness to execute specific behaviour (Klöckner, 2013). The influence of attitude on behavioural intention was confirmed in several previous studies. Foster and Montgomery (2021) found that positive attitude towards behaviour is directly related to behavioural intention. Similarly, studies conducted by Goworek et al. (2017) and Albloushy and Hiller Connell (2019) found that consumer attitudes towards reusable sanitary wear were positively associated with the intention to use them. However, Ahn et al. (2012) found that attitudes do not solely determine intentions and subsequent behaviour. Other factors, such as situational factors, social influence, perceived behavioural control and individual motivations, also play a role in shaping intentions. Xu et al. (2015) further add that consumer attitudes can be a powerful predictor of intention and subsequent behaviour when they are aligned with an individual's values. It is therefore hypothesised that:

H1 Consumer attitudes influence intention to use reusable sanitary towels

Subjective Norms and Intention to Use Reusable Sanitary Towels

Subjective norms measure the degree to which an individual believes that members of an esteemed social group are engaging in the desired or approved behaviours (Khan et al., 2023). According to Cialdini et al. (1990), moral socialisation theory is essential to the formation of subjective descriptive norms. The term subjective norms refers to the shared values that guide individuals to respond to environmental issues or guide performance of pro-environmental behaviour (Chan & Bishop, 2013). Individuals' intention to adopt a specific behaviour was found to be favourably influenced by the belief that it is socially acceptable or widely practiced within their reference group (De Leeuw et al., 2015). Subjective descriptive norms were found to have a positive influence on behavioural intention (Xu et al., 2015; De Leeuw et al., 2015). In an investigation conducted by De Leeuw et al. (2015), descriptive norms were able to account for 29 percent variance in intention. We argue that individuals may start using reusable sanitary towels if they believe that their peers are also doing so. Consistent with the foregoing discussion, it is therefore hypothesised that:

H2 Subjective norms influence intention to use reusable sanitary towels

Perceived Behavioural Control and Intention

Perceived behavioural control measures an individual's view or belief that a specific behaviour may be executed with success (Ahn et al., 2012). It also assesses the degree of control, easiness or difficulties associated with engaging in certain behaviours (Chan & Bishop, 2013). When individuals have high perceptions of behavioural control, they exhibit higher levels of self-efficacy which enhances the formation of favourable behavioural intentions. Conversely, when individuals perceive that there are many barriers that constrain behavioural performance, they are more likely to develop unfavourable intentions (Ajzen, 1991). In the context of reusable sanitary towels, unsupportive social norms and limited availability are identified as the major factors that shape perceptions of behavioural control (Albloushy & Hiller Connell, 2019; Foster & Montgomery, 2021). Based on the foregoing discussion, it is hypothesised that:

H3 Perceived behavioural control influences intention to use reusable sanitary towels

Intention and Actual Behaviour

Intention is regarded as the most immediate predictor of behaviour (Ajzen, 1991). When individuals display favourable intentions, previous studies report that they are more likely to engage in actual behavioural performance (Ghali-Zinoubi & Toukabri, 2019; Su et al., 2022). The link between behaviour and intention was confirmed in previous studies (Armitage & Conner, 2001; Bamberg & Möser, 2007; Klöckner, 2013). However, there is also evidence suggesting that people do not always follow through on their stated goals. For example, a study by Klöckner (2013) revealed that the prevailing situational factors may inhibit actual behavioural performance despite the existence of favourable intentions. Consistent with the forging discussion, it is hypothesised that:

H4 Intention influences actual behaviour of using reusable sanitary towels

Materials and Methods

Research Philosophy

Research philosophy refers to the world views that guide the construction of knowledge in particular field of study (Creswell, 2009). This chapter is guided by positivism. This research philosophy supports the primary objective of this study of examining, using quantitative data, the rational and normative drivers that influence the use of reusable sanitary towels.

Research Design

Research design refers to the overall strategy adopted by the researcher to integrate the components of the study in a consistent, logical and coherent way to ensure effective addressing of the research problem (Saunders et al., 2015). The research design is related to the research philosophy and determines how the research question will be answered. A descriptive research design was utilised in this study as the primary objective was to examine hypothesised relationships between study constructs.

Target Population and Sampling Method

Population refers to the entire group of individuals in the study about which some information is required to ascertain (Boncz, 2015). The characteristics of individuals who are selected from the study population need to be defined prior to the study to avoid confusion and ambiguity (Hu, 2014). Therefore, this study focused on adolescent girls' behavioural acceptance of reusable sanitary towels. The targeted population of study included adolescent girls between the ages of 18 and 34 recruited from high-density informal settlements in Harare, Lusaka and Lilongwe. High-density informal settlements were considered because they are associated with poor living conditions that are tough, which may warrant the need for reusable sanitary wear for young adult women.

Purposive sampling was used to select study participants. This sampling technique was used because it allowed researchers to select respondents who were familiar with reusable sanitary towels. Thus, only adolescent girls between the ages of 18 and 34 with experience using reusable sanitary towels were considered.

Data Collection Instrument and Administration

A structured questionnaire was used to collect data in this study. The questionnaire was developed with a focus on addressing research objectives. Study constructs were measured using validated scales that were adapted from previous studies (Roudposhti et al., 2018; Makudza et al., 2021). To enhance content validity, the questionnaire was pilot tested with a sample of 40 respondents prior to the main survey. With the aid of pilot test results, minor adjustments were effected to the measurement scales of perceived behavioural control and subjective norms. The questionnaires were administered to respondents at community centres responsible for the distribution of free reusable towels to girls in disadvantaged communities. All variables were measured using a five-point Likert scale on a continuum from 1 to 5 (strongly disagree to strongly agree).

Data Analysis

Data were analysed using the Statistical Package for Social Sciences (SPSS) version 25. Descriptive statistics, reliability and validity analysis, correlation and regression analysis were conducted.

Results

Sample Profile

A total of 320 questionnaires were valid for analysis. The majority of the respondents (*n* = 224; 70 percent) were aged between 18 and 22 years. In terms of education level, 62 percent (*n* = 198) were enrolled for a secondary education qualification, 18 percent (*n* = 48) indicated that they dropped out of school and the remaining 23 percent did not disclose their level of education. Out of the 38 percent who were not attending any formal education, 36 percent indicated that they were informally employed.

Reliability Statistics

The study had five variables. The reliability of these variables was tested using the Cronbach Alpha test. Table 5.1 presents the results.

Table 5.1 shows that all variables of the study were reliable. This was proven by Cronbach's coefficients greater than 0.7. According to Shrestha (2021), if the Cronbach Alpha is 0.70 or better, the variable is reliable. In Table 4.2, the subjective norms variable recorded the highest coefficient whilst the actual behaviour variable recorded the lowest of 0.828.

Descriptive Statistics of the Variables

Attitudes Towards Reusable Sanitary Towels

The mean value for the attitude scale was 3.875 out of 5. This overall mean score indicates that respondents agreed that they have a positive attitude towards the use of reusable sanitary towels. This implies that young women in the sample consider the use of reusable sanitary towels as

Table 5.1 Reliability test

	Cronbach's alpha coefficient	*N of items*
Attitudes	0.884	4
Subjective norms	0.897	4
Intention behaviour	0.863	4
Perceived behavioural control	0.850	4
Actual behaviour	0.828	5

Table 5.2 Descriptive statistics for attitude scale

Attitudes towards reusable sanitary towel		*N*	*Min*	*Max*	*Mean*	*Std. Dev*
At1	*To me, using reusable sanity wear is a good practice.*	*320*	*2*	*5*	*3.84*	*0.828*
At2	To me, using reusable sanity wear is likeable.	320	3.0	5.0	3.986	0.7384
At3	To me, using reusable sanity wear is beneficial.	320	2.0	5.0	3.799	0.7807
At4	To me, using reusable sanity wear is wise.	320	2.0	5.0	3.875	0.7744
Summated mean score					3.875	

a good practice, likeable, beneficial and a wise choice. This result is consistent with the findings of previous studies. Kshatriya et al. (2021) found that girls have a positive attitude towards recyclable sanitary pads. However, there are deviations from this study with the results of Kattimani et al. (2024) who found that the urban population were more favourably inclined towards non-reusable sanitary wear, whilst the poor rural population exhibited favourable attitudes. Table 5.2 shows the mean scores for the attitude scale results.

Social Norms Towards Reusable Sanitary Towels

The mean score for social norms was 1.967 which also falls in the agree category. This implies that the prevailing social norms do not support the use of reusable sanitary towels. Table 5.3 presents the scores for the social norm scale.

Perceived Behavioural Control

The mean score for perceived behavioural control was 1930 and this shows that respondents are experiencing challenges when it comes to using reusable sanitary towels. This result also shows that respondents do not have direct control over the behavioural decision to use sanitary towels. Perceived behavioural control is influenced by availability of reusable sanitary towels (mean, 1.456) and inability to make their own decisions (mean, 1.258). Table 5.4 presents the scores for the perceived behavioural control scale.

Table 5.3 Descriptive statistics for social norms

Social norms towards reusable sanitary towels		*N*	*Min*	*Max*	*Mean*	*Std. Dev*
SN1	People who are important to me always use reusable sanitary towels as they make one get stained easily	320	1.0	2.0	2.583	0.8810
SN2	People who are important to me consider use of reusable sanity towels as acceptable	320	1.0	3.0	1.833	0.8446
SN3	I feel socially obliged to use reusable sanitary towels	320	1.0	2.0	1.701	0.8535
SN4	I feel as if I am not obliged to use reusable sanitary wear as they will not be as comfortable as disposable pads	320	2.0	2.0	1.750	0.8236
Summated mean score					1.967	

Table 5.4 Descriptive statistics for perceived behavioural control

Perceived behavioural control		*N*	*Min*	*Max*	*Mean*	*Std. Dev*
PBC1	The decision to use reusable sanitary wear is not completely up to me.	320	1.0	2.0	1.258	0.8518
PBC2	To me, using reusable sanitary wear on a regular basis would not be easy.	320	1.0	2.0	2.424	0.8158
PBC3	Even if I wanted to, I could not easily use reusable sanitary wear.	320	2.0	5.0	2.576	0.8271
PBC4	I find reusable sanitary wear not readily available.	320	2.0	5.0	1.456	0.6954
Summated mean					1.930	

Intention to Use Reusable Sanitary Towels

The mean composite score of intention behaviour was 3.61, which shows a high level of intention among young girls towards using reusable sanitary wear. All questionnaire items which were used to measure intention behaviour had mean scores that were in excess of 3.5, showing high propensity towards the need to use reusable sanitary wear. Thus, it is imperative that the industry is ready to adopt and use reusable sanitary wear. A study by Rozek and Moak (2019) also found that 67 percent of their sample were keen to try out reusable sanitary wear, and that study supports the current findings (Table 5.5).

Table 5.5 Descriptive statistics for behavioural intention scale

Intention to use reusable sanitary towels		*N*	*Min*	*Max*	*Mean*	*Std. Dev*
INT1	I will use reusable sanitary wear in future	144	2.0	5.0	3.632	0.7911
INT2	I make an effort to use reusable sanitary wear	144	2.0	5.0	3.521	0.7931
INT3	I intend to use reusable sanitary wear each time of the month	144	2.0	5.0	3.653	0.7693
INT4	I plan to use reusable sanitary wear monthly	144	2.0	5.0	3.618	0.8528
Mean score intention behaviour					3.61	

Table 5.6 Descriptive statistics for actual behaviour scale

Actual behaviour of using reusable sanitary towels		*N*	*Min*	*Max*	*Mean*	*Std. Dev*
AB1	I use reusable sanitary wear	320	2.0	5.0	1.188	0.8808
AB2	I can use reusable sanitary wear if it is made available for me	320	2.0	5.0	1.211	0.9015
AB3	I use reusable sanitary wear given for free	320	2.0	5.0	1.136	0.8849
AB4	I used reusable sanitary wear after being taught how to make it	320	2.0	5.0	1.292	0.8518
AB5	I prefer reusable menstrual pads to disposable ones	320	2.0	5.0	1.146	0.8402
Summated mean actual behaviour					1.19	

Actual Behaviour of Using Reusable Sanitary Towels

The study found that young women do not use reusable pads. This was proven by a low overall mean score of 1.19 which tended to a disagree category on the measuring instrument. This means that respondents disagreed with the use of reusable sanitary pads. This shows that there is a need for awareness and education towards enhancing adoption of reusable sanitary pads by young women. The study by Scott et al. (2023) had related results, showing that the majority of urban dwellers really preferred disposable sanitary wear to reusable ones (Table 5.6).

Correlation Analysis

The relationship between study variables was examined by computing Pearson Product Moment coefficients. The correlation coefficients ranged from insignificant to significant, as shown in Table 5.7.

Table 5.7 Correlation matrix

Variables	*ATT*	*SN*	*PBC*	*INT*	*AB*
ATT	1.0				
SN	0.307**	1.0			
PBC	−0.065	0.264**	1.0		
INT	0.457**	−0.602**	−0.640**	1.0	
AB	0.438*	−0.523**	−0.581**	0.663**	1.0

Table 5.8 Model 1 results

Coefficients

Model		Unstandardised coefficients		Standardised coefficients	*T*	Sig.	*R*-square
		B	Std. Error	Beta			40.2%
1	(Constant)	1.413	0.359		3.938	0.000	
	Attitudes	0.297	0.076	0.295	3.894	0.000	
	Subjective norms	0.281	0.069	−0.309	−4.076	0.000	
	Perceived behavioural control	0.297	0.066	−0.395	−5.994	0.000	

a. Dependent Variable: Intention

Hypotheses Testing

Multiple regression analysis was used to test study hypotheses. Model 1 included attitude, subjective norms and perceived behavioural control as predictors of intention to use reusable sanitary towels. The model summary in Table 4.5 shows that attitude, subjective norms and perceived behavioural control account for 40.2 percent variance in intention to use reusable sanitary towels. The results showed that attitude has a positive influence on intention to use reusable sanitary towels, while subjective norm and perceived behaviour control were found to have a negative influence on intention to use reusable sanitary towels. Table 5.8 shows Model 1 results.

Model 2 examined the influence of intention on the actual behaviour of using sanitary towels. Intention was found to have a negative effect on the actual behaviour of using sanitary towels. Table 5.9 provides the results of Model 2 assessment.

Table 5.9 Model 2 assessment results

Coefficients						
Model	Unstandardised coefficients		Standardised coefficients	*T*	Sig.	*R*-square
	B	Std. Error	Beta			20.5%
1 (Constant)	1.822	0.321		5.673	0.000	
Intention	0.317	0.083	−0.319	−3.841	0.000	

a. Dependent variable: actual behaviour

Discussion of Results

The first hypothesis predicted that attitude towards reusable sanitary towels positively influences intention to use them. The hypothesised relationship was supported by the data. This result is consistent with findings from previous studies (Goworek et al., 2017; Joseph et al., 2023; Foster & Montgomery, 2021). For example, Joseph et al. (2023) found that attitudes of consumers are a significant driver of intention behaviour and even actual behaviour. Goworek et al. (2017) also found similar results, suggesting that consumer attitudes towards reusable sanitary wear play a significant role in shaping their intention to use them.

The second hypothesis, which predicted a positive influence of subjective norm on intention to use reusable sanitary towels, was not supported by the data. In fact, subjective norms were found to have a negative influence. This result suggests the use of reusable sanitary towels is yet to gain societal approval. This result is consistent with previous studies (Goworek et al., 2017) where the use of reusable sanitary towels was associated with societal disapproval and stigma. Overall, this result shows that social influence plays an important role in shaping respondents' intention to use reusable sanitary towels.

The third hypothesis predicted that perceived behavioural control positively influences intention to use reusable sanitary towels. This proposition was not supported by sample data. In disconfirmation of the proposition, perceived behavioural control was found to have a negative influence on intention to use reusable sanitary towels. This result suggests that respondents surveyed perceive that the decision to use reusable sanitary towels is beyond their own control. This result is supported by findings from previous studies (Albloushy & Hiller Connell, 2019; Foster & Montgomery,

2021) where limited availability of reusable sanitary towels was identified as the major constraining factor.

The final hypotheses predicted that intention has a positive influence on the actual behaviour of using sanitary towels. This proposition was not supported by the data. In fact, intention was found to have a negative influence on the actual behaviour of using reusable sanitary towels. This is despite the fact that respondents in this study reported favourable intentions to use reusable sanitary towels, an indication that individuals do not always act based on their stated intentions. This result is consistent with previous studies that reported the existence of the intention-behaviour gap (Klöckner, 2013; De Leeuw et al., 2015; Ertz et al., 2017). For example, a study by Klöckner (2013) revealed that the prevailing situational factors may inhibit actual behavioural performance despite the existence of favourable intentions.

Implications of the Study

The study found that favourable attitudes lead to increased intention behaviour to use reusable sanitary wear. There is a need to provide educative information about the benefits of using reusable sanitary wear, such as cost savings, environmental impact and health benefits. The government and other private players are thus recommended to raise awareness about the environmental impact of disposable sanitary products. There is also need to offer incentives on reusable sanitary wear. The study recommends marketers and retailers to offer discounts, promotions or free samples of reusable sanitary wear to encourage consumers to try them out. The study also recommends collaborating with influencers to promote a positive attitude towards reusable sanitary wear. This involves partnering with influencers or bloggers who can promote the benefits of reusable sanitary wear to their followers.

The study also found that if social norms are favourable towards the use of reusable sanitary wear, intention to use reusable sanitary wear increases significantly. Community education and awareness campaigns aim to educate young girls and their families about the benefits of reusable sanitary wear, including cost savings, environmental sustainability and health implications. Training and skills development focus on how to use, clean and maintain reusable sanitary wear, as well as how to combat the stigma associated with menstruation. Partnerships with local organisations and collaborations with local NGOs, schools, health centres and community leaders help distribute reusable sanitary wear to young girls in need and promote social inclusivity. Peer-to-peer support networks enable young girls to share their experiences, offer advice and advocate for the use of reusable sanitary wear within their communities.

Improving accessibility of reusable sanitary wear. Reusable sanitary wear should be easily available to consumers by stocking them in stores, providing online purchase options and collaborating with local organisations to distribute them to those in need. Allowing individuals to try out reusable sanitary wear to see how easy and effective they are. The study further recommends hosting workshops or demonstrations to show proper use and care of reusable options. That can drive self-behavioural control. Provision of feedback and encouragement to individuals who are using reusable sanitary wear to reinforce their behaviour is also imperative.

Conclusion

Reusable sanitary towels are positioned as an environmentally friendly and affordable alternative to disposable sanitary towels. The TPB was applied as the underpinning theory. Consumer attitudes, subjective norms and perceived behavioural control were conceptualised as drivers of social innovation. Acceptance of social innovation was premised on intention and actual usage of reusable sanitary towels. A quantitative methodology was employed in which data were collected from 320 young women using a structured questionnaire. The respondents were drawn from Malawi, Zambia and Zimbabwe. Data were analysed using the SPSS Version 25. Subjective norms and perceived behavioural control emerged as the major factors that constrain the use of reusable sanitary towels. The existence of the intention-behaviour gap was confirmed. The results of this study underscore the need to foster social norms and provide facilitating conditions that promote the mainstream use of reusable sanitary towels.

References

Ahn, J. M., Koo, D. M., & Chang, H. S. (2012). Different impacts of normative influences on pro-environmental purchasing behavior explained by differences in individual characteristics. *Journal of Global Scholars of Marketing Science, 22*(2), 163–182.

Ajzen, I. (1991). The theory of planned behavior. *Organizational Behavior and Human Decision Processes, 50*, 179–211.

Albloushy, H., & Hiller Connell, K. Y. (2019). Purchasing environmentally sustainable apparel: The attitudes and intentions of female Kuwaiti consumers. *International Journal of Consumer Studies, 43*(4), 390–401.

Alexandra, C. (2021). *To what extent can social enterprises commercializing environmentally friendly menstrual products improve Menstrual Health and Hygiene in India?* Louvain School of Management, Université catholique de Louvain.

Armitage, C. J., & Conner, M. (2001). Efficacy of the theory of planned behaviour: A meta-analytic review. *British Journal of Social Psychology, 40*(4), 471–499.

Bamberg, S., & Möser, G. (2007). Twenty years after Hines, Hungerford, and Tomera: A new meta-analysis of psycho-social determinants of pro-environmental behaviour. *Journal of Environmental Psychology, 27*(1), 14–25

Bhebhe, S. (2021). *Menstruation management of school girls in rural Zimbabwe. (Master dissertation).* Cape Town: University of South Africa.

Boncz, I. (2015). *Introduction to research methodology.* Faculty of Health Sciences of University of Pécs.

Chan, L., & Bishop, B. (2013). A moral basis for recycling: Extending the theory of planned behaviour. *Journal of environmental psychology, 36*, 96–102.

Cialdini, R. B., Reno, R. R., & Kallgren, C. A. (1990). A focus theory of normative conduct: Recycling the concept of norms to reduce littering in public places. *Journal of personality and social psychology, 58*(6), 10–15.

Creswell, J. W. (2009). Research Design: Qualitative, Quantitative and Mixed Methods Approaches. 3rd ed. Thousand Oaks, CA: Sage.

De Leeuw, A., Valois, P., Ajzen, I., & Schmidt, P. (2015). Using the theory of planned behavior to identify key beliefs underlying pro-environmental behavior in high-school students: Implications for educational interventions. *Journal of environmental psychology, 42*, 128–138.

Ertz, M., Huang, R., Jo, M. S., Karakas, F., & Sarigöllü, E. (2017). From single-use to multi-use: Study of consumers' behavior toward consumption of reusable containers. *Journal of Environmental Management, 193*, 334–344.

Foster, J., & Montgomery, P. (2021). A study of environmentally friendly menstrual absorbents in the context of social change for adolescent girls in low-and middle-income countries. *International Journal of Environmental Research and Public Health, 18*(18), 9766.

Ghali-Zinoubi, Z., & Toukabri, M. (2019). The antecedents of the consumer purchase intention: Sensitivity to price and involvement in organic product: Moderating role of product regional identity. *Trends in Food Science & Technology, 90*, 175–179.

Goworek, H., Hiller, A., Fisher, T., Cooper, T., & Woodward, S. (2017). Consumers' attitudes towards sustainable fashion: Clothing usage and disposal. In *Sustainability in fashion and textiles* (pp. 376–392). Routledge.

Hu, S. (2014). Study population. In *Encyclopedia of quality of life and well-being research.* Springer.

Joseph, J., Elangovan, N., & Sundaravel, E. (2023). Attitude of urban women towards sustainable sanitary products: A study in Bengaluru.

Kambala, C., Chinangwa, A., Chipeta, E., Torondel, B., & Morse, T. (2020). Acceptability of menstrual products interventions for menstrual hygiene management among women and girls in Malawi. *Reprod Health, 17*(185), 1–12.

Kattimani, V., Alekhya, D. L. D., Pathralapati, S., Sojin, S. S., Patel, S., Prabhakar, C., & Dixit, H. (2024). Knowledge, attitude, perception and practices towards disposal of sanitary napkins among young females: A cross-sectional Study. *Journal of Pharmacy and Bioallied Sciences, 16*(Suppl. 1), S637–S640.

Khan, N. A., Hassan, S., Pravdina, N., & Akhtar, M. (2023). Drivers of sustainability: technological and relational factors influencing young consumers' green buying intentions and green actual consumption behavior. *Young Consumers, 24*(6), 686–703.

Klöckner, C. A. (2013). A comprehensive model of the psychology of environmental behaviour—A meta-analysis. *Global environmental change, 23*(5), 1028–1038.

Kshatriya, A., Jani, D., & Patel, V. R. (2021). Influence of consumer's values for attitude towards purchasing and intention towards re–purchasing of sanitary pads among collage girls of Ahmadabad city.

Kumari, K., & Kumar, H. (2023). Menstrual health, grassroots innovations and social entrepreneurship: Exploring the case of Indian "Pad Man". In *20th Anniversary Globelics international conference*. Kerala: Gulati Institute of Finance and Taxation.

Makudza, F., Sandada, M., & Madzikanda, D. (2021). Augmenting social commerce acceptance through an all-inclusive approach to social commerce drivers. Evidence from the hotel industry. *Malaysian Journal of E-Commerce, 5*(2), 55–63.

Mantsebo, T. (2021). *Addressing the menstrual health and hygiene needs of girls and young women*. (Master thesis, Amsterdam (VU): KIT (Royal Tropical Institute)).

Mauye, H. (2024). *Reusable sanitary pads for rural girls*. Retrieved from Hand in hand: https://www.handinhand.co.zw/

Medina-Perucha, L., Lopez-Jimenez, T., Holst, A. S., Jacques-Avino, C., Munros-Feliu, J., Martinez-Bruno, C., et al. (2022). Use and perceptions on reusable and non-reusable menstrual products in Spain: A mixed-methods study. *PLoS One, 17*(3), e0265646.

Mehta, S., Grover, A., Mittal, N., Nanda, P., & Khatuja, R. (2020). Reusable sanitary napkins—Time to revisit. *Journal of Public Health, 44*(2), 356–362.

Nyamayedenga, V. K., & Tsvere, M. (2020). Real time data capture: A response to unsustainable dumping of disposable diapers and sanitary pads in Gweru City, Zimbabwe. *East African Journal of Education and Social Sciences, 1*(2), 54–64.

Nyava-Machadu, T. (2024). *The hidden cost of inaction on menstrual poverty.* Retrieved from Health Times: healthtimes.co.zw/2024/01/09/the-hidden-cost-of-inaction-on-menstrual-poverty/

Onasoga, O. A., Aluko, J. O., Oyekale, R. A., & Mohammed-Ali, A. (2022). Perception and acceptability of reusable menstrual pads among secondary school girls in Ilorin, Kwara State. *Texila International Journal of Nursing, 8*(2), 1–9.

Ramsay, C., Hennegan, J., Douglas, C. H., Eddy, S., Head, A., & Lim, M. S. (2023). Reusable period products: Use and perceptions among young people in Victoria. *Australia. BMC Womens Health, 23*(102).

Roudposhti, V. M., Nilashi, M., Mardani, A., Streimikiene, D., Samad, S., & Ibrahim, O. (2018). A new model for customer purchase intention in e-commerce recommendation agents. *Journal of International Studies, 11*(4).

Rozek, M., & Moak, S. (2019). Promoting Female Empowerment through Reusable Sanitary Pad Project in Namanga, Kenya. In *Proceedings of The National Conference on Undergraduate Research, April* (pp. 663–672).

Sassone, S. C., Silva, S., Metzger, J., Fisher, N., Mohan, A., & Felsman, I. (2022). Reusable sanitary napkins in rural India: A remote quality improvement project for adolescent girls promoting menstrual hygiene health during the COVID-19 pandemic. *IUHPE—Global Health Promotion, 29*(3), 77–85.

Saunders, M. N., Gray, D. E., Tosey, P., & Sadler-Smith, E. (2015). Concepts and theory building. A guide to professional doctorates in business and management, 35–56.

Scott, L., Steinfield, L., Dolan, C., & Dopson, S. (2023). Sanitary pad: Acceptability and sustainability study.

Shannon, A. K., Melendez-Torres, G. J., & Hennegan, J. (2021). How do women and girls experience menstrual health interventions in low-and middle-income countries? Insights from a systematic review and qualitative metasynthesis. *Culture, Health & Summary, 23*(5), 624–643.

Shrestha, N. (2021). Factor analysis as a tool for survey analysis. *American Journal of Applied Mathematics and Statistics, 9*(1), 4–11.

Singh, R., Agarwal, M., Sinha, S., Chaudhary, N., Sinha, H. H., & Anant, M. (2022). Study of adaptability and efficacy of menstrual cups in managing menstrual health and hygeine: A descriptive longitudinal study. *Cureus, 14*(9), e29690.

Su, W., Sun, X., Guo, X., Zhang, W., & Li, G. (2022). An analysis of awe evoked by COVID-19 on green purchasing behavior: A dual-path effect of approach-avoidance motivation. *Frontiers in Psychology, 13*, 952485.

Tembo, M. (2022). *Acceptability, uptake, and effectiveness of a menstrual health intervention among young women in Zimbabwe. (Doctrate thesis)*. School of Hygeiine & Tropical Medicine.

UNHCR. (2019). *Pilot study findings on the provision of hygeine kits with reusable sanitary pads: Testing the appropriateness and acceptability of AFRIpads reusable*. OCHA.
UNICEF. (2019). *The Zimbabwe formative research on menstrual hygiene management (Final Report)*. UNICEF, Government of Zimbabwe.
van Eijk, A. M., Jayasinghe, N., Zulaika, G., Mason, L., Sivakami, M., Unger, H. W., & Phillips-Howard, P. A. (2021). Exploring mentrual products: A systematic review and meta-analysis of reusable menstrual pads for public health internationally. *PLoS ONE, 16*(9), e0257610.
Witte, J. (2021). Enhancing the menstrual experience of menstruating adolescents in Mashonaland Central: A qualitative study. *Glocality, 4*(1), 1–19.
Xu, X., Arpan, L. M., & Chen, C. F. (2015). The moderating role of individual differences in responses to benefit and temporal framing of messages promoting residential energy saving. *Journal of Environmental Psychology, 44*, 95–108.

CHAPTER 6

Efficacy of Green Marketing as a Sustainable Business Model: Multi-Case Study Analysis

Chipo Mukonza and Asphat Muposhi

Introduction

The world is facing unprecedented environmental challenges, from climate change to pollution and resource depletion. As a result, consumers, governments, and businesses are increasingly recognising the importance of sustainability and environmental responsibility. In response, companies are adopting green marketing strategies to reduce environmental footprint, enhance their brand reputation, and drive long-term growth. Sustainable green marketing has emerged as a viable business model, enabling companies to promote eco-friendly products and services, reduce waste, and promote sustainable consumption patterns. This chapter

C. Mukonza (✉)
Loughborough University, Loughborough, UK
e-mail: c.mukonza@lboro.ac.uk

A. Muposhi
Marketing Division, School of Business Sciences,
University of the Witwatersrand, Johannesburg, South Africa
e-mail: asphat.muposhi@wits.ac.za

A. Muposhi, B. Nyagadza (eds.), *Sustainable Green Marketing Strategies for a Circular Economy*, Palgrave Studies of Marketing in Emerging Economies,
https://doi.org/10.1007/978-3-031-81627-7_6

explores the concept of sustainable green marketing as a business model, highlighting its key principles, benefits, and implementation strategies.

Background

The concept of green marketing emerged in the 1970s, but it wasn't until the 1990s that it gained significant attention as a business strategy (Ottman, 2011). Since then, green marketing has evolved to become a key component of corporate social responsibility (CSR) and sustainability initiatives. Today, companies are under increasing pressure to adopt sustainable practices, reduce their environmental impact, and promote eco-friendly products and services. The rise of social media and digital technologies has also created new opportunities for companies to engage with customers, promote sustainable practices, and build brand reputation (Mukonza et al. 2021). In an era marked by increasing environmental awareness and concerns about climate change, the concept of sustainability has moved from the periphery to the centre stage of business discourse. Consumers are increasingly demanding eco-friendly products and services, and companies are recognising the need to integrate sustainable practices into their core business models (Mukonza et al., 2021). Green marketing, a strategic approach that focuses on promoting the environmental benefits of products or services, has emerged as a vital tool for businesses to navigate this shifting landscape. Today's "greener" products (defined as having a lighter impact on the planet than alternatives) and today's more "sustainable" products (those that add a social dimension, e.g., fair trade) now not only work well, but they also likely work better and more efficiently than their "brown" counterparts (Giddings et al., 2024). However, despite the growing importance of green marketing, many companies struggle to implement effective sustainable marketing strategies that drive business growth and profitability.

While green marketing has become increasingly important for businesses, many companies struggle to integrate sustainability into their marketing strategies (Giddings et al., 2024; Chen & Wang, 2023). The lack of clear guidelines, metrics, and frameworks for implementing sustainable green marketing strategies has to some extent hindered companies' ability to effectively promote eco-friendly products and services, reduce waste, and promote sustainable consumption patterns. Furthermore, the increasing scepticism and greenwashing concerns among consumers and stakeholders make it challenging for companies to build trust and credibility around their sustainability claims. Companies often grapple with issues of transparency, credibility, and the need to balance environmental concerns with economic viability. Hence, Chen and Wang (2023) believe that green washing is unintentional and even understandable in a fast-growing industry still finding its sea legs.

Consumers, on the other hand, may struggle to discern genuine green initiatives from mere "greenwashing." Additionally, the long-term implications of green marketing on business performance and environmental sustainability remain a subject of ongoing debate (Mukonza & Swarts, 2020). As a result, companies need a comprehensive understanding of the principles, benefits, and implementation strategies of sustainable green marketing to drive business growth, enhance brand reputation, and contribute to a more sustainable future. This chapter contributes to the evolving debate on green marketing by addressing the following objectives:

- To explore the concept of sustainable green marketing as a business model and its key principles and benefits.
- To evaluate the impact of these strategies on key marketing performance indicators such as brand image, customer loyalty, and market share growth.
- To investigate the roles and influence of various stakeholders, including consumers, employees, and investors, in shaping green marketing strategy.
- To identify and analyse the key metrics and tools used to measure and evaluate the effectiveness of its sustainable green marketing strategies.

This chapter holds significant contributions for both academia and industry. From an academic perspective, it will contribute to the growing body of literature on sustainable marketing and consumer behaviour, shedding light on the complex relationship between green marketing practices and consumer response (Leonidou et al., 2011; Mukonza et al., 2021). Furthermore, it will offer insights into the challenges and opportunities associated with integrating sustainability into business models, thus enriching the discourse on corporate social responsibility and sustainable business practices (Davis, 2022).

From an industry standpoint, the findings of this study provide valuable guidance to companies seeking to implement effective green marketing strategies. Additionally, the chapter aims to help businesses understand consumer expectations, navigate the complexities of green communication, and leverage sustainability as a competitive advantage (Ottman, 2011). Moreover, the chapter's exploration of the economic and environmental benefits of green marketing will encourage companies to embrace sustainability as a core business principle, fostering a more sustainable and responsible marketplace (Mukonza & Swarts, 2020). The chapter explores how sustainable green marketing can drive business growth and profitability, providing valuable insights for companies seeking to align their economic goals with environmental responsibility (Chen & Wang, 2023).

Literature Review

Theoretical Framework

This chapter is anchored in the theoretical concept of corporate social responsibility (CSR), which emphasizes the responsibility of businesses to consider the social and environmental impacts of their operations (Carroll, 1979). Green marketing can be seen as a manifestation of CSR, as it reflects a company's commitment to environmental sustainability and responsible business practices (Maignan & Ferrell, 2004). This, theory for this chapter, emphasises the need for balancing economic, social, and environmental factors in business operations. It provides the foundation for understanding the importance of integrating sustainability into marketing strategies (Davidson, 2022). Drawing from the CSR theory, this chapter proposes a regenerative approach to production and consumption, which emphasises on resource efficiency, waste reduction, and the development of sustainable products.

Green Marketing as a New Business Paradigm

The new green marketing paradigm represents a significant shift in business approaches towards achieving sustainable development and a greener economy (Mukonza & Swarts, 2020). This paradigm, which gained prominence in the late twentieth century, emerged in response to growing public concerns about the sustainability of economic growth and increased environmental awareness. Green marketing encompasses a wide range of activities, including product modifications, changes in production processes, packaging innovations, and advertising adaptations (Singh & Pandey, 2012). The United Nations' Sustainable Development Goals

(SDGs) have played a crucial role in accelerating the adoption of green marketing strategies globally, encouraging businesses to align their practices with sustainability objectives (United Nations, 2015; Giddings et al., 2024).

In this new paradigm, sustainability is no longer an idealistic notion but a key driver of innovation and development (Deloitte, 2023). Companies are increasingly motivated to improve their environmental performance and communicate their sustainability efforts effectively, often through green marketing initiatives, to be included in growing environmental, social, and governance (ESG) stock funds (Davis, 2022). The concept of "greener products" rather than absolutely "green products" has gained traction, acknowledging that every product has potential for improvement (Deloitte, 2023). This shift has made the production and marketing of greener products a business imperative, driven by consumer demand for sustainable products (Deloitte, 2023) and the need to avoid regulatory penalties and reputational damage in an increasingly environmentally conscious marketplace (Rathore, 2018).

The new green marketing paradigm is reshaping the agendas of manufacturers, service companies, government officials, NGOs, and other societal influencers (Ottman, 2011). It emphasises the commitment to developing safe, eco-friendly goods and services, employing recyclable and easily decomposed packaging, implementing better pollution-prevention methods, and using energy more efficiently (Mukonza & Swarts, 2020). As a result, green marketing strategies have become a crucial tool for enhancing corporate image and business performance, particularly in the retail sector, reflecting a fundamental change in how businesses approach sustainability and market their products and services.

However, the term "green marketing" is often associated with "greenwashing," where companies overstate or mislead consumers about their products' environmental attributes (Ottman, 2011). This phenomenon highlights the need for standardised metrics and third-party verification in green marketing claims. Longitudinal studies, such as the GreenBiz and Trucost's *State of Green Business* report (2019), provide valuable insights into the evolution of corporate sustainability practices over time. These reports offer quantitative data on key performance indicators (KPIs) related to environmental impact, resource efficiency, and sustainability integration in business operations.

This varied approach to green marketing and sustainability necessitates further empirical research to quantify the long-term impacts on business performance, consumer behaviour, and environmental outcomes. Future studies should focus on developing standardised metrics for assessing the effectiveness of green marketing strategies and their contribution to overall sustainability goals.

Green Marketing as Business Model

The concept of green marketing as a profitable business model is a complex and evolving area of study, intersecting with various aspects of business strategy, sustainability, and consumer behaviour. This section explores the potential profitability of green marketing business models through several key perspectives. A business model (BM) serves multiple functions, including establishing credibility, defining an organisation's niche in target markets, and demonstrating its position within each ecosystem (Sirohi, 2024). The transformation of traditional BMs into Green Sustainable Business Models (GSBMs) involves incorporating socio-environmental values into operational designs. However, research indicates that such transformations often face challenges and may fail (Leal Filho et al., 2021).

Institutional and technological perspectives have also contributed to the growth of green marketing business models. It stands to reason that global enterprises are increasingly compelled to adopt green marketing business models due to political, legal, and environmental considerations (Bocken et al., 2016). On the other hand, there is the technological perspective which shows that the development of new GSBMs is driven by technological advancements, which can create opportunities for innovation and competitive advantage. Additionally, institutional forces are considered inputs in determining the economic justification for green business model transformations (Stubbs & Cocklin, 2008).

Milton Friedman's 1970 assertion that "the social responsibility of business is to increase profits" relegated environmental issues and corporate social responsibility to the periphery of business concerns. However, this perspective has evolved significantly in recent decades. Research indicates that consumer demand for greener products and services creates opportunities for businesses to (a) promote greener offerings; (b) introduce profitable new products; (c) build top-line sales; (d) enhance corporate image; (e) boost employee morale through engagement in purpose-driven activities (Ottman, 2011). Proactive companies are developing new greener technologies, business models, and designs, which can

(a) capture media attention; (b) attract new customers; (c) establish competitive advantages; (d) potentially alter industry norms and practices.

Deloitte (2023) suggests that (a) the future of business lies at the intersection of global needs, consumer desires, and brand equity capabilities; (b) sustainability issues, like digital technologies, have the potential to fundamentally change existing market mechanisms; (c) green business models often conflict with traditional business practices and structures, necessitating further research to understand effective methods for managing the transformation to green and profitable business models. While there is growing evidence suggesting that environmental sustainability can be profitable, more rigorous, long-term studies are needed to conclusively demonstrate the financial benefits of adopting green business practices across various industries and business sizes.

Materials and Methods

This chapter employed a qualitative case study method to provide an understanding of sustainable green marketing as a business model. This chapter followed a five-staged procedure for conducting case study analysis recommended by Stake (2005). The procedure involved the following steps: determining the appropriateness of a case study, case identification and sampling procedures, developing procedures for data collection, specifying data analysis techniques, and reporting study findings.

A case study approach was selected due to its ability to provide an in-depth analysis of research phenomenon in a specific context using multiple sources of information (Creswell & Poth, 2018). The implementation of green marketing strategies is influenced by a host of organisational and industry factors and for this reason the case study was considered as the most suitable approach. The case studies were selected from South African companies listed on the Johannesburg stock exchange with a focus on those implementing and publicising their green marketing strategies. With the aid of purposeful sampling, Woolworths Holdings, Old Mutual, Pick n Pay, and Nedbank were selected. The case studies were deliberately sampled in order to understand approaches of distinct perspectives in the implementation of green marketing strategies.

Secondary data was obtained from publicly available company reports, sustainability reports, sustainability indices, and financial statements. This data enabled content analysis to be carried out to understand the influence of green marketing strategies on business performance. Ethical considerations were duly observed by ensuring that information obtained is accurate, up-to-date, and from reliable sources. The researcher avoided misrepresenting the company's actions or performance and strived for a balanced analysis.

Brief Overview of Green Marketing Initiatives of Selected Case Studies

Case Study 1: Woolworths Holdings

Woolworths Holdings' commitment to green marketing is reflected in its Good Business Journey strategy. Through this strategy, the company aims to be a leader in responsible retailing in the countries in which it trades. Through its annual Good Business Journey report, Woolworths Holdings has committed to transparency in sustainability reporting. Some of the green marketing initiatives include:

- Replacement of polystyrene plastic for avocado packaging with kraft box base made from 63% recycled paper,
- Marketing of organic products,
- Adopting a plastic-free policy in which the company commits to do away with single-use plastic shopping bags,
- Investment in renewable energy sources, and
- Marketing sustainable fashion (Good Business Journey Report, 2021).

Case Study 2: Old Mutual

- Old Mutual has demonstrated a commitment to sustainability and responsible investment over the past five years, as evidenced by several green products and initiatives.
- Old Mutual offers a range of responsible investment funds that consider environmental, social, and governance (ESG) factors in their investment decisions. These funds provide investors with opportunities to support sustainable businesses and contribute to positive environmental and social impact (Old Mutual, 2022). Old Mutual's Green Bond: In 2020, Old Mutual launched a green bond, which is

a type of investment that is specifically used to fund environmentally friendly projects (Old Mutual, 2020). In addition, Old Mutual offers a range of green insurance products that provide cover for environmentally friendly initiatives, such as solar panels and green buildings (Old Mutual, 2022).

- The organisation has made significant investments in renewable energy projects, such as wind and solar farms, to contribute to the transition to a low-carbon economy. These investments generate clean energy and reduce greenhouse gas emissions. (Old Mutual, 2022). Old Mutual has invested in sustainable agriculture projects, such as macadamia farms in Malawi, that promote sustainable farming practices, support local communities, and generate foreign currency for the economy (Old Mutual, 2022). Old Mutual has implemented green building practices in its own operations and investments, aiming to reduce energy consumption, water usage, and waste generation (Old Mutual, 2022). It has integrated climate risk management into its investment and underwriting processes to assess and mitigate the potential impact of climate change on its business and investments (Old Mutual, 2020). Furthermore, Old Mutual actively engages with policymakers and industry stakeholders to promote responsible investment practices and advocate for policies that support a sustainable future (Old Mutual, 2022). These are just some examples of Old Mutual's green products and initiatives implemented in recent years. The company continues to explore opportunities to enhance its sustainability performance and contribute to a greener future.
- The green initiatives at Old Mutual have contributed to an increase in profits and improved corporate image to a significant extent. For instance, Old Mutual's sustainable investment funds have seen significant growth, with assets under management increasing by 25% in 2020 to R14.4 billion (Old Mutual, 2020). The company's green bond issuance in 2020 raised R2 billion, which was used to fund environmentally friendly projects (Old Mutual, 2020). Old Mutual's energy efficiency initiatives have resulted in cost savings of R12 million per annum (Old Mutual, 2022).

Case Study 3: Pick n Pay

Pick n Pay's Sustainability Strategy is anchored on the 3P Framework focusing on people, profit, and planet. The following are selected green marketing initiatives from Pick n Pay's Sustainability Living Report:

- Promoting health living
- Promoting sustainable packaging
- Marketing organic food
- Investment in green marketing practices such as recycling
- Food waste reduction and
- Ethical sourcing and procurement practices (Sustainability Living Report, 2021)

Case Study 4: Nedbank

Nedbank's Implementation of Sustainable Green Marketing Strategies
Nedbank is deeply committed to sustainable practices, and this is reflected in its green marketing strategies that aim to drive both business growth and environmental responsibility. A snapshot of some of the green marketing strategies are as follows:

- Nedbank aims to have 100% of its lending and investment activity supporting a net-zero carbon economy and is actively developing carbon pathways to guide its transition to net-zero.
- Nedbank offers various green financing products and services that incentivise sustainable practices among its clients (Nedbank CIB, 2023). Some of the examples include green bonds to fund renewable energy and sustainability projects.
- Nedbank has introduced sustainable finance solutions for businesses transitioning to greener operations, also offering green mortgages for energy-efficient homes.
- In line with the Paris Agreement to keep global warming below 2° Celsius, above pre-industrial levels by 2050, Nedbank is transitioning away from involvement in or funding of fossil fuels, while accelerating its efforts to finance renewable energy solutions.
- Nedbank prides itself as the first South African commercial bank to launch a green bond on the JSE; it raised R1.7 billion to fund solar and renewable energy projects, R14 billion for exposures to

green-certified properties, and R11 billion for buildings containing sustainable features. It has also financed a broad range of sustainable projects and initiatives in the energy, property, and agriculture sectors.

- Nedbank employs a multifaceted approach to gauge the success of its green marketing strategies. This includes tracking the financial performance of its green financing products, which serves as a direct indicator of market demand for sustainable solutions (Nedbank CIB, 2023). The bank's numerous accolades, including the 2023 Sustainable Bank of the Year, 2023 Sustainability-linked Loan of the Year (Africa), and 2023 Outstanding Leadership in Sustainable Bonds in Africa, further underscore the efficacy of its long-term commitment to a green business model. These awards are a testament to Nedbank's successful implementation of sustainable practices and its leadership in driving positive environmental change.
- Nedbank measures customer response to its green marketing campaigns through surveys, social media analytics, and feedback mechanisms (Nedbank, 2024).
- Nedbank monitors its environmental performance by tracking metrics such as energy consumption, waste generation, and carbon emissions (Nedbank, 2024).

Results and Discussion

Intersection Between Green Marketing and Corporate Strategy

The analysis of case studies revealed that the convergence of green marketing and corporate strategy represents a complex, multifaceted paradigm in contemporary business models. This intersection is characterised by several key components. Current green marketing trends encompass a range of strategies, including transparency, accountability, product innovation, corporate social responsibility, and stakeholder engagement (Muchenje et al., 2023). These strategies manifest in various forms: (a) cause marketing; (b) green product positioning; (c) eco-labelling and certification; (d) consumer engagement and participation; (e) sustainable supply chain management; (f) transparent reporting; (g) sustainable packaging; and (h) extended product responsibility. For enhanced sustainable marketing, firms are advised to focus on: (a) green marketing initiatives; (b) regenerative practices; (c) social justice and equity considerations; (d) digital sustainability; (e) transparency and accountability measures; and (f) climate change mitigation and adaptation strategies. According to Muchenje et al. (2023) the viability of green marketing as a business model depends on the ability to integrate these strategies seamlessly and align them with the overall corporate strategy.

Green Marketing and Business Growth

The results of case study analyses indicated that investment in green marketing initiatives is associated with business growth. Companies offering green products stand a chance of gaining competitive advantage due to positive consumer attitudes towards environmentally friendly practices. Moreover, sustainable business practices were found to contribute to risk mitigation by reducing exposure to regulatory sanctions, mitigating potential consumer backlash from non-compliance with environmental standards and enhancing resilience to climate-related risks. The link between green marketing and green growth was confirmed by the Carbon Disclosure Project. Figure 6.1 summarises the key benefits of green business models.

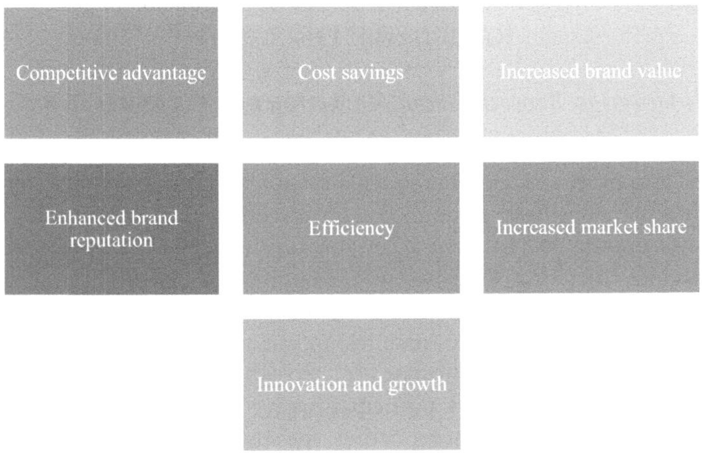

Fig. 6.1 Benefits of a green marketing business model

Green Marketing and Innovation

The case studies analysed showed that the implementation of green marketing initiatives is associated with innovation leadership. From the case studies analysed, innovations were noted in the form of new product development such as green products, carbon credits, green bonds, and carbon emission mitigation techniques. In support of the linkages between green marketing and innovation, Khayyam et al. (2024) posits that investments in technology and collaborations with external partners, guided by strong resource management capabilities, are crucial for achieving sustainable innovation. Efthymiou et al. (2023) noted that it is important for organisations to select the right technology for the right purpose.

Green Marketing Contradictions and Challenges

The case studies analysed and literature reviewed showed that green marketing is a contested terrain. For example, there are mixed views on the effectiveness of green marketing as a strategy for transitioning to a more circular economy. The implementation of green marketing in South Africa was premised on the realisation of circular economy which promotes inclusive growth, environmental sustainability, employment, and social equity (DEA, 2017). However, since the adoption of the green economy vision in 2008, economic and social indicators show an increase in unemployment, slackened economic growth, pollution, and increase in social inequalities (Stats SA, 2021). With a surge in unemployment, which is estimated to be 34% as of September 2024, the realisation of green jobs remains a mirage (Stats SA, 2024).

Green marketing critics also caution that green marketing is evolving into "green capitalism." This concern is emanating from the fact that green marketing in South Africa is mainly implemented under the template of self-regulation. As a result, there is a growing perception that companies are selectively implementing green marketing initiatives that are more aligned towards profitability than environmental sustainability. The majority of organic products are premium priced and beyond the reach of low-income groups, and this is regarded as a new form of market exclusion. Retailers in the grocery sector are being accused of making profit from the sale of reusable shopping bags. A life cycle assessment of some of the reusable shopping bags showed that they are not environmentally friendly as claimed, thereby raising green washing concerns (DEA, 2017).

The case studies analysed revealed that the journey towards a green business model has operational challenges and is characterised by consumer cynicism. They showed that the companies faced challenges during the integration process, often due to factors such as high upfront costs, realignment of organisation culture towards sustainability, lack of internal expertise in formulating and implementing green strategies, and difficulties in accurately measuring environmental impact of value chain processes.

Proposed for Green Marketing Business Model

The desired outcome from the implementation of an integrated green marketing strategy at corporate level should be a viable green business model. A green business model focuses on circular economy principles, such as waste reduction, resource efficiency, and product life extension (Lewandowski, 2016). It incorporates elements of shared value creation, where businesses address societal and environmental challenges while simultaneously generating economic returns (Schaltegger et al., 2016). A green business model encourages businesses to consider the entire life cycle of their products or services, from sourcing and production to distribution and end-of-life disposal, to minimise negative environmental impacts and maximise positive social and economic outcomes (Stubbs & Cocklin, 2008). Furthermore, a green business model framework fosters innovation and encourages the development of new products, services, and processes that contribute to a more sustainable future (Schaltegger et al., 2016). Based on the case studies reviewed, Fig. 6.2 provides the green marketing business model.

Policy Contributions

The findings of this study support the argument for stronger policy interventions promoting sustainable business practices. It provides evidence that green initiatives can enhance financial performance and corporate image, encouraging policymakers to formulate policies accordingly: governments can introduce tax breaks, subsidies, or other incentives to encourage businesses to adopt green initiatives and integrate sustainability into their business models; governments can facilitate the development of green finance mechanisms, such as green bonds and sustainability-linked

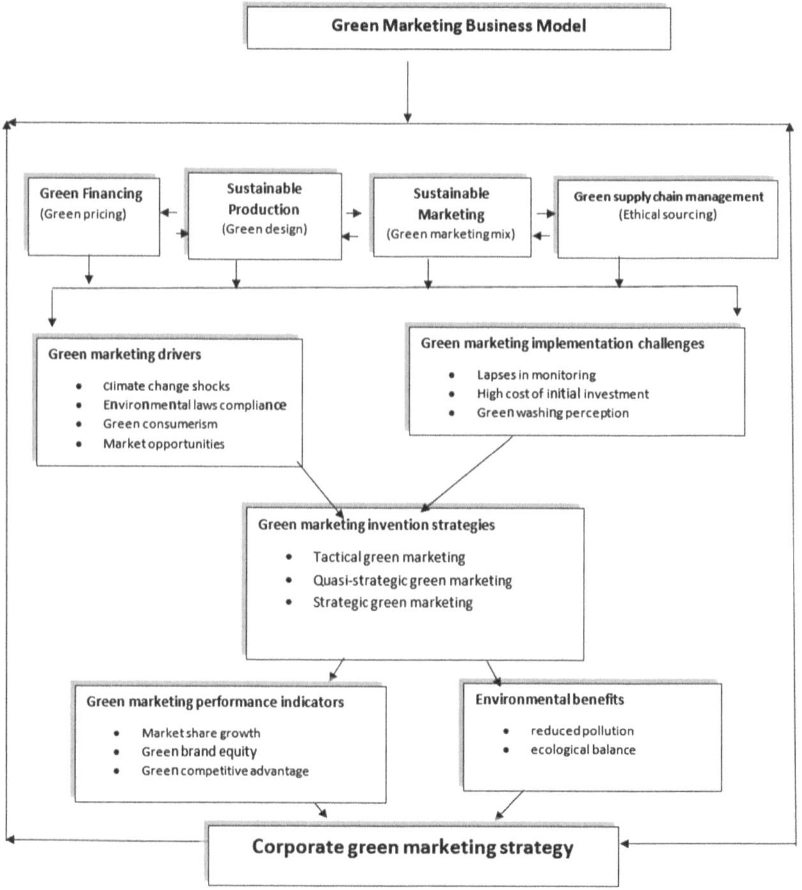

Fig. 6.2 Green marketing business model

loans, to provide companies access to capital for green projects; raising public awareness about the importance of sustainability and the benefits of supporting sustainable businesses can create market demand for green products and services.

The study provides insights into the effectiveness of sustainability initiatives in driving business growth and profitability, which can inform policy decisions related to sustainability and environmental regulations. The study highlights the importance of green marketing as a business model, which can encourage policymakers to develop incentives and regulations that promote sustainable business practices. The study's findings on the growth of sustainable investment funds can inform the development of sustainable investment frameworks and regulations that promote ESG considerations. The study's emphasis on the importance of sustainability in driving business growth and reputation can encourage policymakers to develop regulations and guidelines that promote corporate social responsibility.

Companies should prioritise sustainability and integrate it into their business strategy to drive growth, profitability, and reputation. Companies should invest in green initiatives, such as renewable energy, energy efficiency, and sustainable supply chain practices, to reduce costs and improve their environmental footprint. Companies should develop sustainable products and services that meet the needs of environmentally conscious consumers and investors. Companies should engage with stakeholders, including investors, customers, and employees, to promote their sustainability initiatives and build trust and reputation. Companies should measure and report their sustainability performance using standardised frameworks, such as the Global Reporting Initiative (GRI), to demonstrate transparency and accountability.

Theoretical Contributions

The study extends the concept of green marketing by demonstrating its effectiveness as a business model that drives growth, profitability, and reputation. The study contributes to the theoretical understanding of how sustainability can be integrated into business strategy to drive growth and profitability. The study provides a framework for understanding the key components of sustainable business practices, including green marketing, sustainable supply chain practices, and ESG considerations. The study highlights the importance of stakeholder engagement in promoting sustainability and green marketing, which can inform theoretical models of stakeholder engagement and sustainability.

Conclusion

The findings of this study reveal that green marketing has emerged as a viable business model, enabling companies to reduce their environmental footprint while enhancing brand reputation and driving long-term growth. The case studies analysed showed that commitment to a green marketing strategy has resulted in improved transparency in sustainability reporting, corporate image, cost savings, stakeholder engagement, and responsible investment that considers environmental, social, and governance factors. Greenwashing and high initial capital investment in green marketing technologies were identified as major challenges. The study concludes that green marketing has a potential to be a viable sustainable model as it strategically helps organisations to create sustainable competitive advantage by differentiating themselves by aligning with environmental values and tapping into the market opportunities presented by the growing segment of environmentally conscious consumers.

References

Bocken, N. M., De Pauw, I., Bakker, C., & Van Der Grinten, B. (2016). Product design and business model strategies for a circular economy. *Journal of industrial and production engineering, 33*(5), 308–320.

Carroll, A. B. (1979). A three-dimensional conceptual model of corporate performance. *Academy of Management Review, 4*(4), 497–505.

Chen, H., & Wang, L. (2023). Green marketing and financial performance: A meta-analysis. *Strategic Management Journal, 44*(3), 456–472.

Creswell, J. W., & Poth, C. N. (2018). *Qualitative inquiry and research design: Choosing among five approaches'* (4th ed.). Sage Publications.

Davidson, K. (2022). Sustainability theory in business: A comprehensive review. *Academy of Management Review, 47*(2), 301–320.

Davis, J. A. (2022). Radical Business: How to Transform Your Organization in the Age of Global Crisis. Emerald Publishing Limited.

Department of Environmental Affairs. (2017). Department of Environmental Affairs strives to improve plastic bag recycling in South Africa. Retrieved May 20, 2024, from https://www.environment.gov.za/mediarelease/deaonimproveplasticbagrecyclinginSA

Deloitte. (2023). Many financial services companies mobilize for ESG reporting. https://www2.deloitte.com/us/en/pages/audit/articles/esg-survey/financial-services-companies-sustainabilityreporting.htm

Efthymiou, L., Kulshrestha, A., & Kulshrestha, S. (2023). A study on sustainability and ESG in the service sector in India: Benefits, challenges, and future implications. *Administrative Sciences, 13*, 165.

Giddings, J., Bloomfield, H., James, R., & Blair, M. (2024). The impact of future UK offshore wind farm distribution and climate change on generation performance and variability. *Environmental Research Letters, 19*(6), 064022.

Good Business Journey report. (2021). Woolworths Holdings. https://www.woolworthsholdings.co.za/wpcontent/uploads/2022/09/2022-Good-Business-Journey-Report.pdf

Khayyam, M., Yushi, J., Liu, Q., Idrees, H., Qin, S., & Nurlegul, A. (2024). Leveraging technological readiness and knowledge sources for green innovation: A resource orchestration perspective. *Business Process Management Journal*, Vol. ahead-of-print No. ahead-of-print.

Leal Filho, W., Frankenberger, F., Salvia, A. L., Azeiteiro, U., Alves, F., Castro, P., ... & Ávila, L. V. (2021). A framework for the implementation of the Sustainable Development Goals in university programmes. *Journal of Cleaner Production, 299*, 126915.

Leonidou, L. C., Katsikeas, C. S., & Morgan, N. A. (2011). Greening the marketing mix: Do firms do it and does it pay off? *Journal of the Academy of Marketing Science, 39*(2), 250–270.

Lewandowski, M. (2016). Designing the business models for circular economy—Towards the conceptual framework. *Sustainability, 8*(1), 43.

Maignan, I., & Ferrell, O. C. (2004). Corporate social responsibility and marketing: An integrative framework. *Journal of the Academy of Marketing Science, 32*(1), 3–19.

Muchenje, C., Ruzive, B., Mugoni, E., Katsvairo, H., & Tapera, M. C. (2023). Sustainable Supply Chain Management: Reducing the Environmental Impact of Production and Distribution Processes. Sustainable Marketing, Branding, and Reputation Management: Strategies for a Greener Future, 52–66.

Mukonza, C., & Swarts, I. (2020). The influence of green marketing strategies on business performance and corporate image in the retail sector. *Business strategy and the Environment, 29*(3), 838–845.

Mukonza, C., Hinson, R. E., Adeola, O., Adisa, I., Mogaji, E., & Kirgiz, A. C. (2021). Green marketing: An introduction. Green marketing in emerging markets: Strategic and operational perspectives, 3–14.

Nedbank. (2024). Sustainability at Nedbank. https://www.nedbank.co.za/content/nedbank/desktop/gt/en/aboutus/green-and-caring/sustainability/sustainability-approach.html

Nedbank CIB. (2023). Advancing a net-zero future and green economy. https://cib.nedbank.co.za/sustainability.html

Old Mutual. (2020). Old Mutual's Green Bond. https://www.oldmutual.co.za/investor-relations/old-mutuals-green-bond/

Old Mutual. (2022). Sustainability report 2022. https://www.oldmutual.com/om-docs/blt594158c53915ca5a/Sustainability_Report_2022.pdf

Ottman, J. (2011). *The new rules of green marketing: Strategies, tools, and inspiration for sustainable branding*. Berrett-Koehler Publishers.

Rathore, B. (2018). Green Strategy: Exploring the Intersection of Marketing and Sustainability in the 21st Century. *Eduzone: International Peer Reviewed/Refereed Multidisciplinary Journal, 7*(2), 83-90.

Schaltegger, S., Hansen, E. G., & Lüdeke-Freund, F. (2016). Business models for sustainability: Origins, present research, and future avenues. *Organization & environment, 29*(1), 3–10.

Singh, P. B., & Pandey, K. K. (2012). Green marketing: policies and practices for sustainable development. *Integral Review, 5*(1), 22–30.

Sirohi, R. A. (2024). Corporate power and the rise of intangibles: A study of Indian firms. *Environment and Planning A: Economy and Space, 56*(3), 865–882.

South African Sustainability Awards (SASWA). (2020). Winners. https://www.sasustainabilityawards.co.za/winners/

Stake, R. E. (2005). Qualitative case studies. In N. K. Denzin & Y. S. Lincoln (Eds.), *The SAGE handbook of qualitative research* (3rd ed., pp. 443–466). Sage Publications.

Stats S. A. (2021). Annual report 2021. https://nationalgovernment.co.za/department_annual/437/2022-statisticssouth-africa-(stats-sa)-annual-report.pdf

Stubbs, W., & Cocklin, C. (2008). Conceptualizing a "sustainability business model". *Organization & environment, 21*(2), 103–127.

Sustainability Living Report. (2021). Pick n Pay. https://sustainabilityreports.com/reports/pick-n-pay-stores-limited-2021-sustainable-living-report-pdf/

United Nations. (2015). Transforming our world by 2030: A new agenda for global action Zero. Draft of the outcome document for the UN Summit to adopt the Post-2015 Development Agenda. New York: United Nations.

CHAPTER 7

Examining Consumer Perspectives in Marketing of Green Products

Emmanuel Chao

Introduction

Green marketing has gained a significant research interest in recent years both in consumer-to-consumer and business-to-business markets (Yadav & Pathak, 2016). The history of green marketing can be traced back to the early movements which started around 1960s with a publication of Rachel Carson which was titled "Silent Spring" in 1962. The publication highlighted the risks of pesticides and triggered a widespread public concern about environmental issues. This was later followed by the first Earth Day that took place in 1970, which was a landmark event. From 1980s the term "green marketing" emerged and business firms started to give more attention to consumer interest in green products. From 1990s green marketing evolved both in theory and practice and led to various concepts such as: sustainability, integration, transparency and authenticity,

E. Chao (✉)
Marketing Management Department, Mzumbe University, Morogoro, Tanzania
e-mail: ejchao@mzumbe.ac.tz

A. Muposhi, B. Nyagadza (eds.), *Sustainable Green Marketing Strategies for a Circular Economy*, Palgrave Studies of Marketing in Emerging Economies,
https://doi.org/10.1007/978-3-031-81627-7_7

eco-friendly packaging, local and organic products, green certifications and labels, and several others. As the concept of green marketing evolves, calibrating and achieving customer satisfaction emerged as key factors central to the success of green marketing as a practice.

Previous studies have shown that there is increasing interest in green products, but when it comes to green products the main challenge is bridging the attitude-behaviour gap (Carrington et al., 2014; Yadav & Pathak, 2016). Despite positive attitudes towards green products, many consumers do not consistently purchase them due to perceived higher costs, limited availability, and doubts about the actual environmental benefits (Carrington et al., 2014). However, there are opportunities for businesses to enhance green marketing effectiveness. Singh and Verma (2017) observed that consumers who perceive green products as healthier alternatives show a stronger intention to consume such products. By improving the transparency of environmental benefits, ensuring competitive pricing, and increasing the availability of green products, businesses can better align with consumer expectations and drive sustainable consumption (Prothero et al., 2011).

Status of Green Products in Tanzania

The marketing of green products requires a deep understanding of consumer perspectives and behaviours because consumers are heterogeneous and purchase intention drivers of green products varies. Findings by Lockie et al. (2002) have shown that green products purchase decisions are often driven by several consumer perceptions. By addressing the challenges and leveraging the opportunities in green marketing, businesses can effectively promote sustainable consumption and contribute to environmental protection. The rise of environmental consciousness among consumers has fundamentally reshaped the landscape of marketing. Consumers today are more informed and discerning, demanding transparency and accountability from companies regarding their environmental claims. This shift has been accompanied by advancements in technology, enabling the development of more sustainable products and processes. This chapter explores consumer perspectives that influence green product consumption in Tanzania with the aim of addressing the following research objectives:

- To examine cognitive factors that influence the consumption of green products
- To understand the role of individual and social factors on the consumption of green products

Literature Review

Consumer Perceptions and Behaviour

Understanding consumer perceptions is crucial in green marketing. According to Ajzen (1991) the theory of planned behaviour (TPB) is driven by behavioural motives where the motive is a function of attitudes towards the behaviour, subjective norms, and perceived behavioural control. In the context of green products, consumers' attitudes towards environmental sustainability, perceived benefits, and social influences play critical roles in shaping their purchase intentions (Kalafatis & Pollard, 1999). Perceived quality, trust, and satisfaction for green products were found to be significant predictors of brand equity for green products (Chen, 2010). Similarly, consumer trust in the safety and quality of green products can affect their purchasing decisions (Chen & Chang, 2013).

Several empirical studies have highlighted the importance of consumer perceptions in green marketing. A study conducted by D'Souza et al. (2006) observed that clear communication of the environmental benefits of green products can significantly enhance consumer purchase intentions. Further, consumers who perceive greater individual benefits from green products are more likely to purchase them (Fotopoulos & Krystallis, 2002). Another study by Kahneman and Tversky (1979) on prospect theory revealed that consumers weigh potential losses more heavily than potential gains, which can influence their decision-making process regarding green products.

Green Marketing Consumption Choices

Development in the area of green products has been of strategic relevance in both B2C (Business to Consumer) and B2B (Business to Business) markets. Calibrating and meeting customer satisfaction is the central idea around green marketing. Green products are often referred to as ecological or environmentally friendly. They are designed to conserve environmental resources by incorporating recycled materials or adopting eco-friendly production processes (Suki, 2016). Hsieh (2012) suggested that as global environmental challenges like pollution, climate change and resource depletion continue to escalate, both consumers and professional buyers are increasingly considering green alternatives. Heightened societal awareness of environmental issues, driven by the detrimental effects of industrialization, has pushed companies to integrate sustainability into their core business strategies (Chen, 2010). This shift is further fuelled by pressure from governments, media and various stakeholder groups, compelling firms to address environmental concerns in their operations. Studies have shown that consumers do consider several factors when it comes to making a purchase decision, especially for green products. These factors tend to have a varying degree of weight in such choices. According to Padel and Foster (2005), there is a gap between consumers' positive attitudes towards green products and the actual purchase decision.

Green Marketing Strategies

Sustainability of the environment is among the issues of interest to consumers that companies can address. The author suggests that green consumer products can be introduced by making small changes to current products or removing environmentally harmful features and replacing them with less harmful ones. Another option is to create entirely new green consumer products, which gives a company the opportunity to become competitive.

The choice of strategy often depends on the trade-off between risk and attractiveness (Makower, 2009). There could be risks in creating new green consumer products compared to modifying existing ones. This implies that companies need to consider the market size and consumers' ability to differentiate based on the extent to which the products are green (Ginsberg & Bloom, 2004). Companies with segments that are small in size may benefit by shifting their focus to make their operations greener. Companies that emphasize brand benefits can focus more on greening their products than their processes (Prakash, 2002). Borin et al. (2013) suggested that consumer purchase intentions may differ depending on whether the products are new green products, recycled or those produced by green procedures. Organizations could choose to utilize various combinations such as the least environmentally harmful ingredients, but such a decision should involve a clear communication strategy to convey the green aspects of the products.

Theoretical Framework

This study is underpinned by the theory of planned behaviour and attitude-behaviour-context (ABC) theory. The theory of planned behaviour is rooted from the theory of reasoned action by Ajzen and Fishbein (1980), which explores how personal choice, social environment and other related factors influence the consumer intention (See e.g. Han & Kim, 2010). In the TPB framework, there are three variables, which are attitude, subjective norm and perceived behavioural control. These combine to shape "behavioural intention," which affects actual behaviour. Past studies have expanded this theory in studying consumers' green purchase intentions by adding other new dimensions beyond the original ones. This extension enhances the TPB model's explanatory power by including factors that comprehensively influence these behavioural intentions. The theory serves as an important tool for examining green consumption motive. The addition of new variables that affect these behavioural intentions improves the model's ability to explain.

This theory initially explained that an individual behaviour is driven by their both perceived expectations and behaviour. Behaviour was later recognized that it is not always entirely voluntary or under conscious control and thus they later introduced the concept of "perceived control." This theory suggests that human action is governed by beliefs. These beliefs show personal attitudes or perceptions as to how favourable an action is. Normative beliefs acknowledge that as social beings, individuals are influenced by the opinions and beliefs of others. Control beliefs include factors that facilitate or inhibit the performance of a behaviour, with intentions and subsequent actions being influenced by an individual's belief in their ability to perform specific behaviours. This includes both internal controls, such as self-perceived competence, knowledge, skills and discipline, and external controls.

Guagnano et al. (1995) suggested that the attitude-behaviour-context (ABC) theory is relevant in understanding consumer perspectives on green marketing, specifically on factors that influence the consumer choices for green products (Goh & Balaji, 2016). Integrating the extended theory of planned behaviour (TPB) with the ABC theory provides a better framework in exploring the factors which influence the consumer choices on green products. Factors which have been adopted in this study include product perceived benefits, product features perceptions, price perceptions, and social and environmental perceptions. Past research findings (e.g. Paul et al., 2016; Taufique & Vaithianathan, 2018; Liao et al., 2020) examined the expanded TPB model with a focus on showing the relationship between green purchasing and other variables.

This study suggests to use a theoretical framework based on consumer behaviour theory, incorporating the expanded TPB model and ABC theory to influence consumers' purchase decisions on green products. The proposed framework categorizes factors into two groups:

- Cognitive drivers: such as perceived green benefits and perceived green feature
- Individual characteristics and social factors: encompassing social norms and concerns for environment

Green purchase intention is typically defined as a precursor to actual purchase behaviour. Numerous studies have examined consumers' motive to predict their real behaviour. Intention to purchase green products according to Chen and Chang (2012) is defined as the likelihood that consumers will choose products which are friendly to environment. Consumers often purchase green products to protect or avoid harming the environment. Motive for purchasing green products can be measured using three items: the intention to buy green products, the willingness to switch brands for ecological reasons and the preference for green versions of products. Green purchase intention serves as a critical variable for assessing both current and future customer purchase decisions.

Hypotheses Development

Perceived Green Products Benefits and Features

Previous research findings (Cretu & Brodie, 2007; Elsäßer & Wirtz, 2017) confirmed the association between the quality of the product and consumer satisfaction. According to Ali Khan et al. (2011) green products' quality can be defined by environmental benefits such as ecolabelling, recyclability, energy efficiency, and minimal environmental impact. High-quality green products not only meet these criteria but also contribute to customer satisfaction and loyalty (Tseng & Hung, 2013). Green product quality encompasses features related to energy conservation, pollution prevention and waste recycling (Chang & Fong,

2010). Customers who are knowledgeable about these aspects tend to make purchasing decisions based on the perceived quality of green products, which directly influence their satisfaction and loyalty (Mayer, 2013; Norazah, 2013; Suki, 2016).

Green perceived value can be viewed as consumers' summative evaluation of what they give for and what they receive from a product or service while considering sustainability (Chen & Chang, 2012). Broadly speaking, the green products' perceived value pertains to a summative assessment of the benefits generated from products and services (Kim et al., 2012), encompassing the benefits and utility obtained relative to the time and money invested. Value is often a strong motivator for customers; thus, the product's perceived benefits, as a factor related to product benefits perception, can enhance purchase intentions. According to Zhuang et al. (2010) companies can drive consumer purchase intentions through effectively utilizing the perceived product benefits or value. According to Yaacob and Zakaria (2011) consumers who are environmentally conscious tend to have an inkling to purchase green products on the basis of the benefits to the environment. Product perceived value is an indicator of how customers evaluate the motive to buy. The higher perceived product value of green products correlates with customer buying intentions (Tan & Goh, 2018).

According to Chen and Chang (2013) perceived benefits reflect consumers' evaluation of a brand's environmental excellence. Green products are thought to have more benefits in terms of health attributes compared to conventional products (Krystallis & Chryssoloidis, 2005). The free chemical nature of these products (William & Hammitt, 2001) is one of the key parts of motivation for consumer choice. According to Roitner-Schobesberger et al. (2008) consumers are concerned with residues which are generated after the production of products. Fotopoulos and Krystallis' (2002) research findings have supported the health and safety attributes of the green products. In other words, green products promote both human and environmental safety which in turn reduce health and environmental hazards. Increasingly, consumers are much more concerned with health and sustainability when it comes to agriculture. According to Wandel and Bugge (1997), these two factors are major dimensions when it comes to the consumption of green products, hence we propose:

H1 Perceived product benefits have an influence on the consumption of green products.

Effect of Product Features on Consumption of Green Products

Perceived product features such as quality are key factors influencing consumer satisfaction and behavioural intentions, as they indicate customers' perceptions of product advantages. According to Zeithaml (1988), perceived product features (such as quality) are defined as a customer's overall evaluation regarding the product's superiority compared to other options. Nekmahmud and Fekete-Farkas (2020) suggested that this assessment of product features significantly impacts customer decisions to buy. Product features such as perceived quality are very critical when it comes to consumer choices for products. The perceived green product quality construct was introduced in the theory of planned behaviour by Wang et al. (2020). In the same vein, other studies (Wu & Chen, 2014) discovered that the perceived green product quality positively influences consumer consumption. Product features play an essential role when it comes to defining the product value (Munroe, 2003). Customers often place higher value and are willing to pay a high price premium when there are better product features. Product features often uplift the likability of products due to their uniqueness and attractiveness compared to conventional products (Vindigni et al., 2002). Customer preference for green products influences their purchase motive (Sreen et al., 2018).

H2 Perceived product features have a positive influence on the consumption of green products.

Green Product Price

There has been a growing demand for green products due to increased environmental consciousness. On the other hand, these products are often more expensive due to the costs associated with sustainable practices (Suki, 2016). Customers are often willing to pay high prices for these products when there is a perceived added value, while others may be hesitant because of price concerns (Catoiu et al., 2010). The perceived fairness of the price, where the environmental benefits justify the higher cost, is crucial in influencing purchasing decisions. In other words, the price has an upper hand in making the final decision, thus the combination of price and the social and environmental concerns works jointly in making the final purchase decision for green products. Organizational buyers, in particular, may be reluctant to pay more for green products if they perceive a compromise in features such as quality (Chen et al., 2006). However, many consumers acknowledge that higher prices can be backed up by environmental benefits, resulting in investing more in sustainable products (Chen, 2010; Ranaei Kordshouli et al., 2015). This shift in perspective is also observed among purchasing managers, who may opt to pay higher prices for firms that comply with environmental and social responsibility standards, such as those set by the United Nations Global Compact (Ranaei Kordshouli et al., 2015).

Even though the perceived benefit and product features influence the consumption of green products, price provides a decisive moment. The influence of price on the consumption of green products varies from one consumer to the other and so does the perception of price (Weisstein et al., 2014). High or low price is not absolute for consumers and this is the reason we are using a perceived price. Most consumers tend to have a defined budget when it comes to food consumption. Products which are within the reach of their defined budget and are perceived to be valuable will be consumed. Price perception has a negative influence on product perceived value (Munroe, 2003). On the other hand, when consumers perceive that the price is relatively lower and there is saving advantages in comparison to buying a similar product elsewhere or an immediate substitute, the consumer will often be motivated to make a purchase decision (Grewal et al., 1998). Barbarossa and Pastore (2015) view price as having an upper hand in the influence on purchase decisions. We anticipate the higher price to lower the chances for purchase of green products as opposed to lower prices, thus:

H3 Perceived high price has a negative influence on the consumption of green products.

Social and Environmental Factors

Consumers are heterogeneous and green product purchase motives vary. While the personality variable was added to the theory of planned behaviour (TPB) (Rhodes et al., 2002), there is limited research which has explored social perspectives. Environmental concern, which reflects an individual's awareness of and efforts to address environmental issues, is a critical factor in understanding consumer attitudes towards green products. According to Paul et al. (2016), environmentally conscious consumers tend to have positive attitudes towards green products, leading to a green lifestyle. Environmental concerns significantly influence young consumers' green purchasing decisions (Nekmahmud & Fekete-Farkas, 2020). In the same vein, research by Hartmann and Apaolaza-Ibáñez (2012) found that environmental concerns positively impact attitudes and purchase intentions for green products.

Social factors such as subjective norms, which involve perceived social pressure to perform or avoid certain behaviour (Ajzen, 1991), play an important role in the consumer's decision to purchase green products. Subjective norms positively impact the consumption motive for green products (Bong Ko & Jin, 2017). When individuals believe that others support green behaviour, they are more likely to adopt them (Yeon Kim & Chung, 2011). The social perspectives such as environmental friendliness, ecological sustainability, wellbeing and related social reasons are increasingly becoming influential in making consumption decisions, especially for middle and higher income groups. Environmental sustainability is one of the SGDs (United Nations sustainable development goals) that promote the consumers to be socially and environmentally conscious. Chinnici et al. (2002) noted the changes in consumption patterns towards the preference for products which have less environmental impact. In other words, ecological behaviour has been witnessed to influence the tendency of consumers towards green products (Harper & Makatouni, 2002). Consumers will act and think as same as their reference group no matter whether it comes from a negative or positive drive. Pickett-Baker and Ozaki (2008) emphasized this through their finding which indicated that social influences promote behaviour changes in a community or group using an example of recycling activities.

H4 Social perceptions have a positive influence on the consumption of green products.

H5 Environmental perceptions have a positive influence on the consumption of green products.

Materials and Methods

Selected Product Category

The research centred on organically grown fruits and vegetables, which possess both desirable and undesirable quality attributes, such as colour, appearance, flavour (taste and aroma), texture and nutritional value. Product appearance plays a crucial role in consumer acceptance or rejection, while nutritional value, a "hidden" attribute is increasingly valued by consumers, as well as by scientists and medical professionals (Kramer, 1965). These products were selected because previous research has primarily focused on green purchasing models using high-involvement products like automobiles, while low-involvement products that are frequently purchased have been largely overlooked (Gupta & Ogden, 2009; Barbarossa & Pastore, 2015).

Measurement Scales

The study variables were operationalized using validated scales that were adapted from previous similar studies. Minor modifications were made to ensure that the study context was taken into account. Consumer buying decision was measured using a 4-item scale that was developed by Wee et al. (2014). Safety perceptions were measured using a 4-item scale that was adapted from Michaelidou and Hassan (2008). A 3-item scale developed by Vindigni et al. (2002) was used to measure quality perceptions. Health perceptions associated with green products were measured using a 4-item scale that was developed by Krystallis and Chryssoloidis (2005). A 3-item scale which was developed by Munroe (2003) was used to measure price perceptions associated with green products. Lastly, environmental perceptions were measured using a 3-item scale that was adapted from Wee et al. (2014). Prior to the main survey, the questionnaire was pilot-tested with a sample of 40 respondents and the reliability of the measurement scales was confirmed. All variables were operationalized using a 5-point Likert scale.

Data Collection

This study examined the impact of consumer perceptions on the purchasing behaviour of green food products in Tanzania. Data was collected for a period of three months using a structured questionnaire from consumers who buy green/organic products in five local markets in Tanzania. A total of 650 questionnaires were distributed, and 399 were valid for analysis, which translated to a response rate of 61.2%.

Data Analysis

Data analysis was conducted using SPSS version 26. Descriptive statistics, correlation analysis (see Table 7.2) and multiple regression analysis (Table 7.3).

Reliability Assessment

The reliability of measurement scales was assessed by computing the Cronbach's alpha coefficient. The Cronbach's alpha coefficients for all variables were above the recommended values. Table 7.1 provides reliability assessment results.

Table 7.1 Reliability test

Construct	*Cronbach's alpha*	*Number of used items*
Health perceptions	0.755	4
Safety perceptions	0.904	4
Quality perceptions	0.768	3
Price perceptions	0.821	3
Envnt perceptions	0.772	3
Actual buying decision	0.880	4

Table 7.2 Correlations

	1	*2*	*3*	*4*	*5*
1.Product benefits perceptions	1				
2.Product features perceptions	0.507**	1			
3.Price perceptions	0.755**	0.491**	1		
4.Social & envnt perceptions	−0.137*	0.118	−0.252**	1	
5.Consumer purchase decision	0.665**	0.485**	0.550**	−0.004	1
Mean	**3.03**	**2.89**	**3.21**	**3.42**	**3.60**
SD	**0.84**	**1.01**	**1.15**	**1.05**	**1.01**

*Correlation is significant at the 0.05 level (2-tailed)

**Correlation is significant at the 0.01 level (2-tailed)

Results

A multiple regression analysis was conducted to examine the consumers' personal perceptions about product benefits, product features, price, and social perceptions. Product benefits, product features, price and social perceptions were the independent variables. Table 7.3 below shows the summarized findings from the analysis. The results have revealed that product benefits, product features, price and social and environmental perceptions to be significantly and positively related to the actual buying decision. The model below has Adj. R-square of 0.468, meaning that the model can explain about 46.8% of the total variation in the actual purchase decision that is explained by the independent variables. Table 7.3 and 7.4 below provides the detailed information on the results.

Discussion of Results

H1 suggested that perceived product benefits have a positive impact on the consumption of green products. This hypothesis was supported by the data ($\beta = 448$, $t = 3.926$, $p < 0.01$). This finding is consistent with previous studies, such as those by Liao et al. (2020) who revealed that purchase decisions of green products are often driven by the perception that green

Table 7.3 Multiple regression analysis

	Unstandardized coefficients		*Standardized coefficients*	*T*	*Sig.*
	B	*Std. Error*	*Beta*		
(Constant)	0.275	0.303		0.907	0.366
Perceived product benefits	0.448	0.114	0.340	3.926	0.000
Perceived product features	0.376	0.110	0.342	3.424	0.001
Price perceptions	−0.202	0.079	−0.209	−2.560	0.011
Social perceptions	−198	0.072	−203	−1.802	070
Environment perceptions	0.086	0.057	0.082	1.506	0.134
R^2				0.480	
Adj. R^2				0.468	
***F* value**				38.05	

Table 7.4 Hypothesis summary

S/N	*Hypothesis*	*Code*	*Findings*
1	Perceived health from GFPs will positively influence the actual buying decision	H_1	Supported
3	Perceived product features of GFPs will positively influence the actual buying decision	H_2	Supported
4	Price perception of green products will negatively influence the actual buying decision	H_3	Supported
5	Social perception has a positive influence on consumption of green products	H_4	Not supported

products do not have artificial additives (such as chemicals, or pesticide residues), and are better for their health. This result also aligns with research findings, who observed that health consciousness is a key attribute that shape consumers' attitudes, intentions and decisions regarding the purchase of green food products.

Similarly, Roitner-Schobesberger et al. (2008) demonstrated that health-related benefits significantly and positively affect customers' intentions and decisions to buy green products. In essence, the greater the perceived health benefits of green products, the stronger the intention and likelihood of consumers actually purchasing them (Taufique & Vaithianathan, 2018; Liao et al., 2020). The demand for green products has increased due to changes in consumer preferences towards food safety and nutritional benefits. Moreover, Yadav and Pathak (2016) found that health benefits are a primary motivator for consumers when choosing organic products, reinforcing the importance of perceived product benefits in influencing the consumption of green products.

H2 posited that product features have a positive influence on the consumption of green products. This hypothesis was supported by the data ($\beta = 376$, $t = 3.4$, $p < 0.01$). This result is consistent with that reported in previous studies (Aschemann-Witzel & Zielke, 2017; Nekmahmud & Fekete-Farkas, 2020). Consumers' perceptions of product features, particularly quality, play a significant role in influencing their actual purchasing decisions regarding green products (Nekmahmud & Fekete-Farkas, 2020). This implies that the stronger the consumers' perceptions of the features of green products, the more likely they are to decide to consume them. Additionally, more recent research by Aschemann-Witzel and Zielke (2017) suggests that consumers' perceptions of the sensory attributes of green products, such as taste and texture, are crucial in determining their purchasing decisions. These sensory attributes, alongside perceived product quality, are often seen as indicators of the overall value of green products, further reinforcing the positive impact of product features on consumption.

H3 proposed that price perception negatively influences the consumption of green products. This hypothesis was supported ($\beta = -202$, $t = -2.5$, $p < 0.01$). This finding is aligned with Shafi and Madhavaiah's (2013) study wherein price perceptions significantly affect consumer decisions regarding the consumption of green products. This result is further supported by the prospect theory, which posits that consumers are sensitive to losses compared to gains (Kahneman & Tversky, 1979). The research findings concur with this theory, as a negative significant relationship was observed between the price paid and the attitude towards the purchase of green products. However, it is important to note that, as reported by Xie et al. (2020), the perception of higher quality associated with green products can sometimes mitigate the negative impact of price, especially among consumers with strong environmental values.

H4 suggested that social perceptions have a positive impact on the consumption of green products. The hypothesis was not supported by the data ($\beta = 066$, $t = 1.5$, $p < 0.05$). This result suggests that consumers tend to prioritize their own personal benefits over social concerns when it comes to consuming green products. H5 suggested that environment perceptions have a positive influence on consumption of green products. The hypothesis was not supported by the data ($\beta = 086$, $t = 1.5$, $p < 0.05$). This result is consistent with previous studies (Ko & Jin, 2017). This indicates that marketers may need to adopt alternative strategies for promoting green products, focusing on aspects other than environmental benefits to achieve successful sales.

Study Implications

The study revealed that perceived product benefits are the key influence on consumer purchasing decisions regarding green food products (GFPs). Perceived product features also positively impacted consumer behaviour, while social considerations were not a major concern among the study population. This suggests that marketers should avoid emphasizing social benefits when promoting GFPs, as these are not deemed relevant by the consumers surveyed. Instead, marketers are encouraged to focus on the product benefits of GFPs, as these are more likely to align with consumers who value these attributes.

The lack of significant influence on social factors in the consumption decision for the green products call for those agencies which deal with promoting the social courses such as environmental protection to redefine their campaigns by either enhancing the value proposition or to combine these campaigns with product benefits and features.

Theoretical Implications

There were few perspectives from the extended theory of planned behaviour which were adopted from this study. Other variables from this theory of planned behaviour and related theories of consumer behaviour in green product can address the gaps in this area. Variables related to personal preferences and consumer beliefs are essential in purchasing decisions; thus, including these variables could enhance the predictive capacity of the model which was used in this study.

Future Research

Future research in this area offers significant opportunities for expanding both theoretical and practical understanding in this area. Researchers could explore the different categories of GFPs, such as grains and cereals, or investigate other green product segments, including health and beauty products, consumer durables, paints and furnishings, paper and cleaning products. Furthermore, future studies could expand the geographical scope to enhance the generalizability of results by employing similar conceptual models and constructs. Researchers may also consider including constructs like GFP taste and consumer beliefs, as suggested in Ajzen's extended behaviour theory. Culture plays an influential role when it comes to the way that consumer use products. Future studies can also extend the focus in examining the area of culture. There is increasing changes in lifestyle which are attributed by globalization. When one compares youths of generation Z, they have different ways on how they make choices. Future researches may seek to explore these trends and demographic changes and how they influence the consumption of green product.

Limitations

Several limitations need to be taken into consideration when interpreting the findings of this study. The study focused exclusively on green fruits and vegetables, which means the results may not be generalized to all green or organically produced products. Further, the study primarily examined one dominant sales channel (local markets), with a focus only on supermarkets. Given the varying characteristics of different channels such as open markets consumer buying behaviours may differ significantly across these contexts. The availability of GFPs is also highly dependent on demand levels. Data collection was also limited to Tanzania, which restricts the generalizability of the findings to other countries. The sampling method used was convenience sampling. The use of convenience sampling may not accurately represent the study population.

Conclusion

This chapter examines consumer perspectives on green marketing with a focus on factors that drive green products consumption decisions in Tanzania. Developing economies present a unique research setting due to differences in cultural and economic factors as compared to developed economies. This chapter utilizes cross-sectional quantitative data which was collected from a sample of 399 respondents. The results show that intention of consumers to purchase green products is driven by their perceived product benefits, perceived product quality and price perceptions. The influence of social and environmental concerns was found to be insignificant. The findings of this study contributes towards efforts directed at effectively positioning green products in developing economies.

References

Ajzen, I. (1991). The theory of planned behavior. *Organizational Behavior and Human Decision Processes, 50*(2), 179–211.

Ajzen, I., & Fishbein, M. (1980). *Understanding attitudes and predicting social behavior*. Prentice-Hall.

Ali Khan, M., Ahmed, S., & Shahzad, A. (2011). Environmental benefits and customer satisfaction through eco-labeling. *Journal of Cleaner Production, 19*(9–10), 1172–1181.

Aschemann-Witzel, J., & Zielke, S. (2017). Can't buy me green? A review of consumer perceptions of and behavior toward the price of organic food. *Journal of Consumer Affairs, 51*(1), 211–251.

Barbarossa, C., & Pastore, A. (2015). Why environmentally conscious consumers do not purchase green products. *Qualitative Market Research: An International Journal, 18*(2), 188–209.

Bong Ko, S., & Jin, B. (2017). Predictors of purchase intention toward green apparel products: A cross-cultural investigation in the USA and China. *Journal of Fashion Marketing and Management: An International Journal, 21*(1), 70–87.

Borin, N., Mulkin, M., & Krishnan, R. (2013). Consumer perceptions of environmentally friendly products. *Journal of Consumer Marketing, 30*(7), 508–520.

Carrington, M. J., Neville, B. A., & Whitwell, G. J. (2014). Lost in translation: Exploring the ethical consumer intention-behavior gap. *Journal of Business Research, 67*(1), 2759–2767.

Catoiu, I., Vrânceanu, D., & Filip, A. (2010). Green marketing strategies: Advantages and challenges. *Amfiteatru Economic Journal, 12*(27), 207–221.

Chang, N. J., & Fong, C. M. (2010). Green product quality, green corporate image, green customer satisfaction, and green customer loyalty. *African journal of business management, 4*(13), 2836.

Chen, Y.-S. (2010). The drivers of green brand equity: Green brand image, green satisfaction, and green trust. *Journal of Business Ethics, 93*(2), 307–319.

Chen, Y.-S., & Chang, C.-H. (2012). Enhance green purchase intentions: The roles of green perceived value, green perceived risk, and green trust. *Management Decision, 50*(3), 502–520.

Chen, Y.-S., & Chang, C.-H. (2013). Towards green trust: The influences of green perceived quality, green perceived risk, and green satisfaction. *Management Decision, 51*(1), 63–82.

Chen, Y.-S., Lai, S.-B., & Wen, C.-T. (2006). The influence of green innovation performance on corporate advantage in Taiwan. *Journal of Business Ethics, 67*(4), 331–339.

Chinnici, G., D'Amico, M., & Pecorino, B. (2002). A multivariate statistical analysis on the consumers of organic products. *British Food Journal, 104*(3/4/5), 187–199.

Cretu, A. E., & Brodie, R. J. (2007). The influence of brand image and company reputation where manufacturers market to small firms: A customer value perspective. *Industrial Marketing Management, 36*(2), 230–240.

D'Souza, C., Taghian, M., Lamb, P., & Peretiatkos, R. (2006). Green products and corporate strategy: an empirical investigation. *Society and business review, 1*(2), 144–157.

Elsäßer, M., & Wirtz, B. W. (2017). Rational and emotional factors of customer satisfaction and brand loyalty in a business-to-business setting. *Journal of Business & Industrial Marketing, 32*(1), 138–152.

Fotopoulos, C., & Krystallis, A. (2002). Organic product avoidance: Reasons for rejection and potential buyers' identification in a countrywide survey. *British Food Journal, 104*(3/4/5), 233–260.

Ginsberg, J. M., & Bloom, P. N. (2004). Choosing the right green marketing strategy. *MIT Sloan Management Review, 46*(1), 79–84.

Goh, S. K., & Balaji, M. S. (2016). Linking green skepticism to green purchase behavior. *Journal of Cleaner Production, 131*, 629–638.

Grewal, D., Monroe, K. B., & Krishnan, R. (1998). The effects of price-comparison advertising on buyers' perceptions of acquisition value, transaction value, and behavioral intentions. *Journal of Marketing, 62*(2), 46–59.

Guagnano, G. A., Stern, P. C., & Dietz, T. (1995). Influences on attitude-behavior relationships: A natural experiment with curbside recycling. *Environment and Behavior, 27*(5), 699–718.

Gupta, S., & Ogden, D. T. (2009). To buy or not to buy? A social dilemma perspective on green buying. *Journal of Consumer Marketing, 26*(6), 376–391.

Han, H., & Kim, Y. (2010). An investigation of green hotel customers' decision formation: Developing an extended model of the theory of planned behavior. *International Journal of Hospitality Management, 29*(4), 659–668.

Harper, G. C., & Makatouni, A. (2002). Consumer perception of organic food production and farm animal welfare. *British Food Journal, 104*(3/4/5), 287–299.

Hartmann, P., & Apaolaza-Ibáñez, V. (2012). Consumer attitude and purchase intention toward green energy brands: The roles of psychological benefits and environmental concern. *Journal of business Research, 65*(9), 1254–1263.

Hsieh, Y.-C. (2012). Hotel companies' environmental policies and practices: A content analysis of their web pages. *International Journal of Contemporary Hospitality Management, 24*(1), 97–121.

Kahneman, D., & Tversky, A. (1979). Prospect Theory: An Analysis of Decision under Risk. *Econometrica, 47*(2), 263–91. https://doi.org/10.2307/1914185

Kalafatis, S. P., & Pollard, M. (1999). Green marketing and Ajzen's theory of planned behaviour: A cross-market examination. *Journal of Consumer Marketing, 16*(5), 441–460.

Kim, H.-Y., Lee, Y., & Yoo, J. (2012). The role of perceived consumer effectiveness and motives on willingness to pay for eco-labeled products. *Management Decision, 50*(7), 1131–1150.

Kramer, A. (1965). Methods for quality evaluation of fruits and vegetables. *Journal of Food Quality, 1*(1), 15–25.

Krystallis, A., & Chryssoloidis, G. (2005). Consumers' willingness to pay for organic food: Factors that affect it and variation per organic product type. *British Food Journal, 107*(5), 320–343.

Liao, S. H., Chen, C. C., Wu, C. F., & Tseng, C. M. (2020). Green consumption behavior in Japan: An examination of relationships between perceptions of green products, costs, and benefits. *Sustainability, 12*(7), 2876.

Lockie, S., Lyons, K., Lawrence, G., & Mummery, K. (2002). Eating 'green': Motivations behind organic food consumption in Australia. *Sociologia Ruralis, 42*(1), 23–40.

Makower, J. (2009). *Strategies for the green economy: Opportunities and challenges in the new world of business*. McGraw-Hill.

Mayer, M. C. (2013). The role of perceived value in the link between quality and willingness to pay for organic food. *Agricultural Economics Review, 14*(1), 21–31.

Michaelidou, N., & Hassan, L. M. (2008). The role of health consciousness, food safety concern and ethical identity on attitudes and intentions towards organic food. *International journal of consumer studies, 32*(2), 163–170.

Munroe, K. B. (2003). *Pricing: Making profitable decisions* (3rd ed.). McGraw-Hill.

Nekmahmud, M., & Fekete-Farkas, M. (2020). Why not green marketing? Determinates and challenges of green marketing adoption in firms' business strategy: A review of literature. *Sustainability, 12*(22), 9587.

Norazah, M. S. (2013). Green purchase intention: The antecedents of green behavioral intention among Malaysian consumers. *Procedia Social and Behavioral Sciences, 130*, 44–45.

Padel, S., & Foster, C. (2005). Exploring the gap between attitudes and behaviour: Understanding why consumers buy or do not buy organic food. *British Food Journal, 107*(8), 606–625.

Paul, J., Modi, A., & Patel, J. (2016). Predicting green product consumption using theory of planned behavior and reasoned action. *Journal of Retailing and Consumer Services, 29*, 123–134.

Pickett-Baker, J., & Ozaki, R. (2008). Pro-environmental products: marketing influence on consumer purchase decision. *Journal of consumer marketing, 25*(5), 281–293.

Prakash, A. (2002). Green marketing, public policy and managerial strategies. *Business Strategy and the Environment, 11*(5), 285–297.

Prothero, A., Dobscha, S., Freund, J., Kilbourne, W. E., Luchs, M. G., Ozanne, L. K., & Thøgersen, J. (2011). Sustainable consumption: Opportunities for consumer research and public policy. *Journal of Public Policy & Marketing, 30*(1), 31–38.

Ranaei Kordshouli, H., Ebrahimi, P., & Allahyari Bouzanjani, A. A. (2015). An analysis of the factors affecting the consumer's intention to purchase green products with the mediation role of the green brand image. *Journal of Consumer Marketing, 32*(7), 555–565.

Rhodes, R. E., Courneya, K. S., & Hayduk, L. A. (2002). Does personality moderate the theory of planned behavior in the exercise domain?. *Journal of Sport and Exercise Psychology, 24*(2), 120–132.

Roitner-Schobesberger, B., Darnhofer, I., Somsook, S., & Vogl, C. R. (2008). Consumer perceptions of organic foods in Bangkok. *Thailand. Food Policy, 33*(2), 112–121.

Shafi, S., & Madhavaiah, C. (2013). Online shopping: Customer satisfaction and loyalty. *Asian Journal of Research in Business Economics and Management, 3*(11), 90–106.

Singh, A., & Verma, P. (2017). Factors influencing Indian consumers' actual buying behaviour towards organic food products. *Journal of Cleaner Production, 167*, 473–483.

Sreen, N., Purbey, S., & Sadarangani, P. (2018). Impact of culture, behavior, and gender on green purchase intention. *Journal of Retailing and Consumer Services, 41*, 177–189.

Suki, N. M. (2016). Green product purchase intention: Impact of green brands, attitude, and knowledge. *British Food Journal, 118*(12), 2893–2910.

Tan, M. L., & Goh, Y. N. (2018). The influence of perceived value on green purchase intention. *Journal of Asian Finance, Economics, and Business, 5*(4), 227–234.

Taufique, K. M. R., & Vaithianathan, S. (2018). A fresh look at understanding green consumer behavior among young urban Indian consumers through the lens of theory of planned behaviour. *Journal of Cleaner Production, 183*, 46–55.

Tseng, M. L., & Hung, S. W. (2013). A strategic decision-making model considering the social costs of carbon dioxide emissions for sustainable supply chain management. *Journal of Environmental Management, 127*, 102–110.

Tversky, A., & Kahneman, D. (1988). Rational choice and the framing of decisions. Decision making: Descriptive, normative, and prescriptive interactions, 167–192.

Vindigni, G., Janssen, M. A., & Jager, W. (2002). Organic food consumption: A multi-theoretical framework of consumer decision making. *British Food Journal, 104*(8), 624–642.

Wandel, M., & Bugge, A. (1997). Environmental concern in consumer evaluation of food quality. *Food Quality and Preference, 8*(1), 19–26.

Wang, Y., Wiegerinck, V., Krikke, H., & Zhang, H. (2020). Understanding the purchase intention towards remanufactured products: A mixed method study. *Journal of Cleaner Production, 257*, 120–122.

Wee, C. S., Ariff, M. S. B. M., Zakuan, N., Tajudin, M. N. M., Ismail, K., & Ishak, N. (2014). Consumers perception, purchase intention and actual purchase behavior of organic food products. *Review of integrative business and economics research, 3*(2), 378.

Weisstein, F. L., Asgari, M., & Siew, S. (2014). Price perceptions and consumer shopping behavior: A field study. *Journal of Business Research, 67*(1), 1689–1696.

William, M., & Hammitt, J. K. (2001). Valuing the health effects of air pollution. *Environmental Science & Technology, 35*(4), 590–596.
Wu, S. I., & Chen, Y. J. (2014). A consumer purchasing model with neural networks: Application of artificial intelligence technology in decision making. *Journal of Business Research, 67*(5), 523–530.
Xie, B., Wang, L., Yang, H., Wang, Z., & Zhang, M. (2020). Price competition in green supply chains: An integrated approach. *International Journal of Production Economics, 223*, 107–120.
Yaacob, M. R., & Zakaria, M. N. (2011). Green product and green purchase behavior: The moderating effect of price sensitivity. *International Journal of Business and Management, 6*(7), 105–116.
Yadav, R., & Pathak, G. S. (2016). Intention to purchase organic food among young consumers: Evidences from a developing nation. *Appetite, 96*, 122–128.
Yeon Kim, H., & Chung, J. E. (2011). Consumer purchase intention for organic personal care products. *Journal of consumer Marketing, 28*(1), 40–47.
Zeithaml, V. A. (1988). Consumer perceptions of price, quality, and value: A means-end model and synthesis of evidence. *Journal of Marketing, 52*(3), 2–22.
Zhuang, W., Luo, X., & Rojas, R. R. (2010). The impact of perceived value on behavior intention: An empirical study. *Journal of Business Research, 63*(8), 820–826.

CHAPTER 8

Fostering Sustainable Tourism Through Green Marketing Practices at World Heritage Tourism Sites

Zibanai Zhou and Tendai Chibaya

Introduction

While tourism is glorified worldwide for its economic contribution, there are growing concerns on sustainability challenges in this industry (Shereni & Saarinen, 2020). Fragile world heritage sites (WHS) are facing an unprecedented crisis of over-tourism, triggering multiple challenges for tourist destination managers (Madhushika & Sivesan, 2022). Over-tourism is increasingly emerging as a plague impairing otherwise iconic

Z. Zhou (✉)
Department of Tourism, Hospitality & Leisure Sciences, Faculty of Business Sciences, Midlands State University, Gweru, Zimbabwe
e-mail: zhouz@staff.msu.ac.zw

T. Chibaya
Department of Hospitality, Tourism and Culture, Great Zimbabwe University, Masvingo, Zimbabwe
e-mail: tchibaya@gzu.ac.zw

A. Muposhi, B. Nyagadza (eds.), *Sustainable Green Marketing Strategies for a Circular Economy*, Palgrave Studies of Marketing in Emerging Economies,
https://doi.org/10.1007/978-3-031-81627-7_8

tourist sites (Zhou & Makwara, 2024). To magnify its negative consequences, over-tourism threatens to reverse the benefits of tourism, strain resources at local level, stoke tourism resentment by local communities and extinction of tourist sites. Unsustainable tourism practices create a lot of challenges on tourist sites. They have suffered environmental degradation, overcrowding, lost original allure and stature (Zhou, 2018; Mkono, 2019). However, the situation is not hopeless. One way to address over-tourism and widespread commercialisation of tourist sites is embracing sustainable green marketing practices. These are responsible approaches, which are conscious of the potentially negative effects caused by over-tourism. WHS are fragile tourist centres, hence exceeding their capacity tolerance levels further exposes their vulnerability. To what extent are sustainable green marketing practices a panacea to negative challenges triggered by over-tourism at world heritage sites? The chapter aimed to examine sustainable green marketing practices at WHS in Zimbabwe.

World heritage tourist sites are regarded as tourist drawcards in most destinations provided; they are well managed and designated. Inclusion of the sites on the prestigious global heritage listing supports tourism development in many ways (Madhushika & Sivesan, 2022). The global community recognises the role played by world heritage sites because of the vast universal value derived from them, and Zimbabwe is not an exception. UNESCO has identified five world heritage sites in Zimbabwe namely Mana Pools and Victoria Falls/Mosi-oa-Tunya categorised as natural while Great Zimbabwe Monuments (1986), Khami Ruins National Monument (1986) and Matobo hills (2003) were listed under the cultural category (Siyabona Africa, 2024). UNESCO listed Great Zimbabwe Monuments in 1986, Khami Ruins National Monument in 1986 and Matobo Hills in 2003 under the UNESCO World Heritage Sites.

Background

In general terms, lack of sound waste management practices and cleaning facilities have been argued to contribute towards the destruction of the pristine environment. Utami et al. (2022) observed a general trend whereby tourists are increasingly attracted and prefer to visit clean tourist sites. Tourists prefer to visit world heritage sites with unspoiled environments. This was supported by Nekmahmud and Fekete-Farkas (2021) who outlined that most tourists are more conscious of environmental issues and opt for environmentally friendly services. This calls for sustainable green marketing or environmental marketing strategies as a way to achieve green tourism from the tourism supply side. The sustainability of the environment can be practised through green tourism. Green tourism refers to the principles of practising good environmental management by tourism destination authorities in and around cultural and heritage resources (Lee et al., 2016). This maintains the aesthetic value of the environment and preserves cultural heritage resource base intact. There are multiple benefits associated with practising green tourism, including reducing adverse effects associated with tourism. To this end, it is important to protect the natural resources in and around the two world heritage sites to ensure that present and future tourists benefit equally. This perspective aligns with the Sustainable Development Goal number 12, of the United Nations 2030, which focuses on responsible consumption and production. In support, Joshi and Rahman (2015) outlined that the significant contribution of WHS on economic development has heightened an urgent need to protect such sites from being destroyed by irresponsible tourism.

Regarding the protection of the world heritage tourist sites, the greening of the environment is vital for sustainable tourism development. The United Nations World Tourism Organisation [UNWTO] and United Nations Development Programme [UNDP] (2018) asserts that proper management of tourism goes a long way in protecting cultural heritage and natural assets. Worldwide, WHS are drivers of tourism, which generate trade opportunities, empower host communities and foster peace and intercultural understanding. However, a growing number of domestic and international tourists travelling each year exert pressure on world heritage sites, straining resources. This has impacted local communities and cultural assets negatively through increased economic leakages, and gas emissions (UNWTO UNDP, 2018). According to Handriana and Ambara (2016) and Rahbar and Wahid (2011) excessive tourism activities cause environmental degradation and destruction of natural resources. In support of this perspective.

Sustainable tourism is at the centre of the UN 2030 SDGs (sustainable development goals). In Zimbabwe, sustainable tourism practices as a subject of research has been predominantly confined to hotels at the expense of world heritage sites. Despite evidence of growing challenges faced by such sites, no single study has examined sustainable marketing practices at world heritage sites in Zimbabwe. Notable studies around the theme of green tourism in Zimbabwe are acknowledged. Zengeni et al. (2013)

assessed hoteliers' perception on green tourism impacts on operating costs of hotels. Nyahunzvi (2014) examined greening of Gweru urban lodges. Njerekai (2019) assessed the level of adoption of demand-oriented hotel green practices in Zimbabwe. Dube and Nhamo (2021) assessed emissions of greenhouse gases and sustainability in Victoria Falls focusing on tour operators, hotels and other related attractions. Shereni and Rogerson (2022) examined sustainable practices and associated benefits in the hotel sector in Zimbabwe. Muposhi et al. (2024) examined customers' willingness to pay and patronage intention using green hotels in Victoria Falls. Broadly, these studies are restricted to hotels with no reference to sustainability issues peculiar to WHS. Overall, there remains limited studies that examined world heritage sites' specific sustainable marketing practices. This study seeks to address this literature gap basing on two world heritage sites in Zimbabwe. In spite of the lofty significance of WHS in the tourism sector in Zimbabwe, Victoria Falls and Mana Pools have rarely been subjects of scholarly analysis in the context of sustainable marketing practices. World heritage sites (WHS) are also influenced by a diverse range of stakeholders at various levels; for example, global, supranational, national and local (Schmutz & Elliott, 2016). The non-governmental organisations (NGOs) have become a significant voice in WHS issues (Elliott & Schmutz, 2012), influencing environmentalism (Shandra et al., 2012), and sustainable tourism (Delgado-Torres & Polomeque, 2012).

Green tourism is an area of interest from academics, tourism practitioners and organisations concerned with protection of finite resources. As already outlined, significant studies have been carried out in Zimbabwe on the theme of green tourism; however, their major weakness is that they are remotely connected to WHS. Most studies have predominantly focused on the hotel sector, on specific aspects of energy efficiencies, tour operators and waste disposal. Thirdly, the marketing perspective remains underrepresented. Evidence of tourism massification and surge in negative impacts spawned by tourism activities around WHS sites is an issue of significant concern in most emerging countries. This is a growing trend worldwide, and what is disturbing most is the reality that most developing countries are at risk given their low and limited adaptive capacities. To this end, WHS are saddled with a plethora of challenges.

On a wider global context, a similar trend is noticed whereby research on green marketing remains dominant in hotels (Nekmahmud & Fekete-Farkas, 2021). Gryshchenko et al. (2022) found out that green tourism enterprises have the opportunity to obtain green certification if they improve their recreational facilities in Ukraine. Kinoti (2011) examined green marketing intervention strategies and sustainable development. The study found that green marketing can potentially solve environmental challenges faced by tourist sites. However, literature around world heritage sites that deals with sustainable green marketing practices in developing tourist destinations remains scanty. It is worth to mention that the discourse around sustainability of WHS and utilisation of suitable marketing practices is incipient. There is evolving debate on the significance of listing a site on the world heritage list (WHL) for inspiring/stimulating the motivation to visit such sites (Cellini, 2011; Cuccia et al., 2013; Poria et al., 2013). However, Mitova et al. (2021) bemoaned paucity of studies around sustainable green marketing practices, natural heritage sites and protected areas nexus. There remain limited studies around the marketing practices at natural heritage sites (Lai & Shafer, 2005; Sangpikul, 2010; Sadiq & Adil, 2020). Studies specifically focusing on sustainable marketing strategies employed at WHS in Zimbabwe in the digital environment are lacking. The present study addresses this lacuna. To this end, the study aligns with evolving research in tourism and green tourism practices; however, heritage tourism at WHS remains underrepresented. Therefore, this study responds to environmental challenges faced by world heritage tourist sites amid increasing calls for the protection of the environment.

This is against a backdrop of growing concern about over-tourism, over-marketing and an increase in negative impacts of tourism at mega tourist sites driven by the need to maximise tourist inflows and tourism investment with little regard to long-term socio-cultural, environmental and economic impacts. Therefore, the chapter addresses these research questions. What are the sustainable green marketing practices adopted by world heritage tourist sites in Zimbabwe? Do sustainable green marketing practices lead to increased tourist satisfaction and loyalty, and do they contribute to the well-being of local communities? What are the main challenges faced by tourism operators in implementing sustainable green marketing practices at Victoria Falls and Mana Pools? What opportunities exist for enhancing the effectiveness of sustainable green marketing practices at these sites?

Evidence of global environmental challenges is overwhelming on WHS, hence justifying the adoption of sustainable green marketing practice. This chapter has huge practical significance value aimed to protect WHS and inspire sustainable tourism development. It is also significant as it adds literature on underresearched sustainable marketing practices at WHS in

Zimbabwe. The study will inspire custodians of WHS to review and adopt effective sustainable marketing practice to protect the sites. The chapter proceeds with an analysis of dominant perspectives around sustainable green marketing practices in a circular economy, and conceptualisation of sustainable tourism development around WHS. The methodology, presentation and discussion of findings follow subsequently, and ends with a conclusion. The next section presents a theoretical framework which underpins the study.

Theoretical Framework Development

The study is underpinned by three concepts namely the circular economy, sustainable tourism development and sustainable green marketing practices. The three concepts are connected by the common goal of achieving sustainability. Each one is briefly explained below.

Circular Economy

The concept of a circular economy refers to an economic system based on the reuse and regeneration of products and resources in an environmentally friendly way. The notion of a circular economy has gained traction worldwide with businesses seeking ways to minimise waste and enhance efficiency. The tourism sector is no exception. There are four main pillars which underpin a circular economy, namely: reuse—extending product life cycles, recycling—closing the loop, waste reduction—designing efficiency and product redesign—rethinking the system. A circular economy aims to achieve economic prosperity and environmental quality. It is depicted as a combination of reduce, reuse and recycle (Kirchherr, et al., 2017). A circular economy describes an economic system that is based on business models which replaces the 'end-of-life' concept with reducing, alternatively using, recycling and recovering materials in production, distribution and consumption process (Kirchherr, et al., 2017). Circular economy is regarded as an effective tool for triggering a sustainable development process. Circular economy is applicable to any resources, including WHS. A circular economy has been posited as a solution to environmental decimation with growing global economic prosperity, by introducing new systems of production, consumption and disposal (Chamberlin & Boks, 2018). It is concerned about responsible production and consumption patterns. Consumption of resources has compromised the planet's regenerative capacity, hence the circular economy seeks to be the panacea (Aguiler-Morales et al., 2023). The circular economy as a paradigm seeks more responsible production and consumption patterns. It replaces the concept of life cycle from 'cradle to grave' with reduction, reuse, recycle, as well as recovery of materials for the benefit of current and future generations (Negrete-Cardoso et al., 2022). It speaks of the real challenges grappling with business firms in the twenty-first century.

The Theory of Sustainable Tourism Development

Sustainable tourism development (STD) is a multi-dimensional concept which includes environmental protection, economic development, social equality and ecological effect. It improves the quality of life at each destination. Sustainable tourism development refers to the sustainable use of resources based on tourism-quality criteria, the principles of sustainable development at the destination and codes of conduct to achieve sustainability (Gonzales, 1996). Thus, sustainable tourism development refers to a grounded process which encourages the global management of resources to ensure their viability, enabling natural and cultural capital and protected areas to be preserved. It prioritises protection of human dignity, emphasises the ethical and social responsibility of both tourists and the local community. Notable attributes in sustainable tourism development are supporting factors and resources, core resources and attractions, destination management and planning. Tourism development plans must be compatible with not only market needs but also human and environment integrity to maintain its economic viability (Hassan, 2000). Sustainable tourism advocates for the co-existence of the natural and physical environment with economic, socio-cultural and health and safety objectives of localities and nations. Sustainable tourism development remains a universal goal for any tourism destination worldwide. Sustainable tourism is anchored on three pillars, namely economic, environmental and socio-cultural pillars. Its overarching objective is to protect the environment from being damaged by overuse. How are sustainable green marketing practices incorporated and advance STD at WHS? The benefits of sustainable tourism development in the context of WHS encompass resource conservation/protection, improved long-term profits at WHS, improved quality of tourist experience as there is no overcrowding, limited litter, good quality environment, clean environment, improved tourist/customer satisfaction, improved customer reviews/ratings, enhance repeat business, enable tourism service providers to charge more hence boosting revenue, favour regeneration of resources by inflicting marginal harm as tourist numbers are controlled, reduced negative impacts of tourism activities and increase in tourist inflows.

Sustainable Green Marketing Practices

Sustainable green marketing practices broadly refer to applying good practices related to environmental care—green labelling, eco-efficiency, reuse of material, and green advertising (Alharthey, 2019). Green marketing is also identified as a strategic policy that companies can adopt to increase their profitability through sustainable management that seeks to meet the needs of environmentally concerned customers (Pardana et al., 2019). Green marketing is the union between the satisfaction needs, the generation of profitability and the concern for helping the environment (Sedky & AbdelRaheem, 2021). It came into being because of the need by firms to attract and retain customers (Liao et al., 2020). Green marketing is predicated on four pillars, namely product development—products must be eco-friendly from the choice of raw materials, design, marketing, consumption and disposal.

Market development—carrying out business with ecological responsibility and with a minimum negative impact on the natural environment. Giving orientation towards satisfying the needs of customers who have environmentally responsible purchasing behaviour. Green communication—involving advertising strategies aimed to improve tourists' attitudes towards a green brand and reinforce their purchasing behaviour towards green products. Green marketing practices—contributing towards good health, well-being and customer brand loyalty (Amoako et al., 2020). Sustainable development is the ability to meet the needs of present generation without compromising the ability of future generations to meet their own needs. Both sustainable development and green marketing are aligned to the same objectives: based on satisfying the needs and preserving the environment. The most applied green marketing practice is green advertising, that is, use of ecolabels (Atkinson & Rosenthal, 2014). Green marketing practices—use of ecolabels, use of eco-friendly plastic bags. Green marketing practices focused on the development of eco-friendly products protect the environment through energy conservation, resource optimisation, reuse of components, recycling, reduction and the elimination of toxic agents.

Literature Review

The Link Between Sustainable Green Marketing and World Heritage Tourism Sites

Natural world heritage sites were threatened with natural destruction as well as changes in social and economic activities (UNESCO, 2021). Park et al. assert that ecological and social challenges play an important role in shaping the kind of problems service providers must address which include climate variation, natural resource management and social conflict. In response to the challenges faced by the world tourism business community, tourism service providers adopted a triple-bottom-line approach that takes into consideration the environment, profit and people (Kumar et al., 2012). This implies that tourism service providers should not only consider the success of tourism at WHS through revenues obtained but also consider tourism's contribution to the community and the environment. Broadly, sustainable tourism development safeguards the interests of future tourists. This premise was amplified by the Brundtland Commission report of 1987. The report essentially highlighted the notion of sustainability as development which is futuristic and friendly to the environment. It ensures that the present generation benefits from the resources without jeopardising future generations from benefiting from the same resources and meets the needs and wants of the present generation without compromising the needs of tomorrow's generations. In essence, environmental, socio-cultural and economic issues should work in harmony within the discipline of sustainable tourism development. According to Fonsera et al. (2020) development should be relevant to the present generations and safeguard future generations to satisfy their own needs. Visitor satisfaction at WHS is regarded as an indication of sustainable tourism (Gidey & Sharma, 2017). However, sustainable heritage tourism development is only possible provided tourists are satisfied with the tourism product offering (Asmelash & Kumar, 2019). Zheng et al. (2020) emphasised that local communities play a significant role in ensuring sustainable tourism development. It is widely acknowledged that listing of sites on the world heritage list is a tool of green marketing strategy by tourism service providers. This combines environmental protection and tourism development. This also forms part of environmental sustainability, a major pillar of sustainable tourism. Tourist sites designated as world heritage sites underpin tourism development trends, influence development of a diversified tourism product and inspire overall tourism growth.

National governments and local development initiatives often pursue inscription of sites on the UNESCO World Heritage List because of its performance in attracting international tourism. Su and Lin (2014) and Yang and Lin (2010) are of the view that countries are inspired to have tourist sites listed on the WHL because of its capability to attract tourist inflows. In addition, it also impacts favourably in terms of image and brand. To this end, many countries aspire their tourist sites to be included in the list in order to get the many benefits associated with the listing (Adie, 2017; King & Halpenny, 2014; Ryan & Silvanto, 2014; Wuepper & Patry, 2017). World heritage tourism service providers can leverage on the inclusion of sites on the world heritage list and practise green marketing approaches for long-term benefits coupled with environmental protection. Environmental protection is also another selling point in the tourism industry (Utami et al., 2022). World heritage tourism service providers are encouraged to practise sustainable marketing as compared to conventional marketing. Conventional marketing is criticised for its focus on increasing revenue without taking into consideration the people and the environment. Sustainable marketing is a combination of social marketing, green marketing and environmental marketing. Sustainable, green and environmental or ecological marketing are words used synonymously to show that the marketing strategies implemented are long-term and focus on protecting the environment from damage (Utami et al., 2022; Madhushika & Sivesan, 2022). The period between 1990 and 2000 saw the emergence of sustainable green marketing issues into the limelight (Jayakumar, 2017). In the tourism and hospitality domain, the term 'green' is mostly preferred. Green tourism is divided into four categories, namely: (a) environmental preservation in order to optimally balance the ecosystem, (b) respect and appreciate diverse culture, (c) promote the local economy, and (d) enrichment of tourists' experience by combining human, physical environment and natural sites (Dodds & Jopp, 2001). The world heritage site within the category of natural environment is credited for its product greening elements. Such natural features lure tourists who are conscious of protecting the environment. Sustainable green marketing is a strategy that promotes the environment as well as the organisation by producing environmentally and socially acceptable goods and services with eco-friendly features. This implies that the green strategies are tailor-made to mitigate the negative effects of tourist activities on the environment.

More so, Cavallo (2015) added that green marketing is a strategic mix-up of different components to be managed such as: matching green demand and green offering, having the knowledge to operate in the eco-market, research and development of competencies for the promotion of green products and services, managerial plans to reduce the environmental effects and maximise various opportunities of the green market and ensure that the environmental goals are achieved. World heritage tourism service providers also need to pay particular attention to environmental issues through managing the carrying capacities of the natural heritage sites. It also calls for tourism service providers to monitor the effects of tourist waste, and emissions on the surrounding environment as a solution to improve the relationship between world heritage sites, environment and tourism business. Thus, tourism activities should satisfy the tourism needs of visitors and the community without jeopardising the environment (Polonsky, 1994).

Another green marketing strategy employed by different stakeholders is the use of ecolabels. Environmentally friendly tourism products are recognised (D'Souza et al., 2006; Proto et al., 2007). Besides using ecolabels, organisations promote their products and services in an environmentally friendly way through the use of Internet. This is part of green marketing communication. The Internet or e-marketing allows service providers to market their products and services in an environmentally friendly way.

These green marketing strategies are influential marketing tactics that boost traveller's green purchasing behaviour. Ottman et al. (2006) opine that green marketing offers benefits such as an increase in market share or profits, increased traveller satisfaction, improved quality of the environment, enhanced competitive advantage and promote the protection of the environment. Overall, green travellers are regarded as protectors of the environment and sources of revenue. This is in line with a prominent doctrine in tourism which underscored a need for visitors to bear in mind that they have to take nothing but pictures or photos and also leave nothing but footprints as they visit natural world heritage sites. The next section looks at world heritage sites.

World Heritage Sites

Research around WHS has attracted scholarly attention (Zhang et al., 2023). Conservation of world heritage sites has gained momentum worldwide (Shi et al., 2019; Kenterelidou & Galatsopoulou, 2021). WHS offer multiple benefits in terms of image and branding (Kim et al., 2019; Wang & Yuan, 2020). However, WHS continue to face challenges posed by the external environment and others attributed to tourism. These undermine the authenticity and originality of such sites and sustainable efforts; natural disasters and tourism activities threaten the integrity and sustainability of such sites (Margottini & Vilimek, 2024; IUCN, 2017). Tourism development at WHS is a double-edged sword which can cause damage to landscape's aesthetic value and cause ecological threats while promoting economic growth. Zhang et al. (2023) outline that tourism development around WHS can have both favourable and unfavourable impacts. If uncontrolled, it destroys the environment. On the contrary, if sustainably developed it can protect the environment. Consequently, adopting sustainable green marketing practices at WHS is hugely beneficial for protecting the environment and fostering sustainable tourism. From an ecological perspective, WHS are fragile and vulnerable to human-attributed actions (He et al., 2021). Furthermore, irrational development and utilisation, coupled with huge tourist numbers, and a fragmented tourism development approach can cause a lot of damages. Adoption of sustainable marketing practices is therefore argued to inspire the development of and protection of the integrity of the environment. In theory, sustainability entrenches the protection of the environment ahead of profits at WHS. WHS are characterised as universal goods, hence must be protected for the long-term benefit of humankind for generations and generations (Conradin & Wiesmann, 2014). Victoria Falls and Mana Pools are categorised as iconic global heritage sites similar to Table Mountain in South Africa, Mount Kilimanjaro in Tanzania, Okavango Delta in Botswana, Niagara Falls in the US, Great Wall of China, Rift Valley in East Africa and Coral reefs in Australia. These sites continue to receive an influx of tourists from across the globe, leading to environmental, socioeconomic and service quality challenges. There is also tremendous pressure from politicians and local communities who also seek greater involvement and access to benefits from the exploitation of such resources. The custodian of such tourist sites are implored to strike a balance among a flurry of stakeholders' competing interest. Against this backdrop, it thus becomes crucial to put in place sustainable practices aimed to preserve the sites. Overcrowding, deteriorating aesthetic value, over-tourism, strained provision of sanitation, climate change issues, over-commercialisation, overdevelopment, change in land use, inclusive tourism, local community empowerment and land degradation are issues of growing concern (Huang et al., 2019). Sustainable or green tourism came into the limelight as part of the twenty-first-century tourism lexicon in 2000; with the culmination of the millennium development goals (MDGs) the United Nations 2030 sustainable development goals (SDGs) have evolved. Schmutz and Elliott (2016) observed a disturbing trend of concerning issues of tourism and sustainability, which WHS grapple with. The study laid bare regional variations in terms of threats posed to WHS by external factors. Traditionally, tourist destination authorities at WHS are preoccupied by the euphoria associated with the listing of cultural heritage sites. More often, the authorities pay little regard to the huge responsibility that comes with such listing. A responsibility to protect such resources from destruction (Schmutz & Elliott, 2018). Such concerns resolve around negative social, economic and environmental impacts induced by tourism on cultural and natural properties (Huang et al., 2019). The study argued that the commoditisation or universalisation of heritage tourism significantly imperils WHS. Tourism at WHS is also argued to have positive impacts. It provides justification for the preservation of ecological resources and touristic value of such sites, create jobs and supports livelihoods for local communities. The next section examines eco-labelling.

3.2 Eco-labelling ISO (2000) defines eco-labelling as a tool which gives product information relating to its character in the environment. Tourists' influx cause overcrowding and pollution issues which are inconsistent with sustainable tourism development at WHS. Eco-labelling provides entities with competitive advantage. Ecolabels are framed as transactional environmental governance tools. Concern has also been raised around excessive emission of greenhouse gases (GHG) connected to the forecast increase in air travel worldwide in the foreseeable future (Samuels & Platts, 2020).

Eco-branding identifies the products which are not harmful to the environment and society. It gives firms a comparative advantage in a highly competitive marketplace. Rahbar and Wahid (2011) outline variants of eco-branding as encompassing symbols, names, designs and terms, among others. Merrrilees (2016) further argues that eco-branding is associated with nature conservation. There are multiple motivations for eco-labelling by tourist sites/organisations. These include greening tourism supply, corporate social responsibility (CSR), attracting new markets and receiving higher prices, among others. These are framed as business benefits and green philosophy (Dunk et al., 2016; Yilmaz et al., 2019). Eco-labelling provides a marketing advantage. Juvan and Dolnicar (2014) opine that tourists' heightened level of awareness of potential environmental crises caused by tourism activities at WHS can be observed. There is a noticeable behavioural change in terms of tourist attitude on the negative impacts of tourism on fragile tourist sites. Yilmaz et al. (2019) outline that ecolabels result in a higher sense of customer loyalty and that ecolabel certificate improves the quality of the product. Ecolabels communicate an organisation's attitude to the environment. To this end, eco-labelling can improve customer perceptions (Espron et al., 2014). There are bright future prospects as research suggests that a segment of business tourists is comfortable paying a 10 per cent markup to stay in 'green' accommodation facilities or visit an eco-friendly tourist facility (Ban et al., 2016). This gives credence to the view that ecolabels enable entities to charge higher prices. This is also supported by Jurado-Rivas and Sanchez-Rivero (2019), who found that some visitors were prepared to pay extra for sustainable products, as is the case in Spain's world heritage city of Caceres. Yilmaz et al. (2019) outline that environmentally friendly hotels have strong linkages with local economy, and are concerned with preserving traditional culture. Tourists are increasingly more concerned with environmental issues (Yttredal & Homlong, 2019). However, ecolabels have their downside largely related to costs of certification and recertification and mandatory annual fees. The next section examines green washing.

Greenwashing of Products and Businesses

Greenwashing is an indiscriminate sharing of negative information of an organisation to present a favourable image of the organisation in the marketplace. Font et al. (2017) frame greenhushing as the intentional partial disclosure of information to an organisation's stakeholders concerning its environmental sustainability practices. The practice of greenwashing is detrimental to an organisation. It deflates customer confidence as symbolised by the eco-labelling and eco-marketing efforts. Victoria Falls and Mana Pools can adopt ecolabels for national park authorities and the International Standardisation Organisation (ISO) certification. Victoria Falls and Mana Pools can use the Biosphere Responsible Tourism ecolabel and Green Globe ecolabel as part of sustainable marketing practices.

Green certification is mainly concerned about a harmonious relationship among the environment-, nature- and local society-friendly and sustainable development. From a socio-ecological viewpoint, there is a strong connection between recreational tourism, perception of ecosystem services and demand for tourism services (Ruskule et al., 2018). According to Ecolabel Index (2019) an excess of 50 ecolabels exist in European countries, while 'Nature's Best' is the most prestigious tourism ecolabel in Scandinavian countries. All nature tourism products are certified. Acampora et al. (2018) outlined that eco-certification of recreational tourism facilities encourages users of such facilities to go green. To this end, we see the proliferation of ecolabels aimed at alleviating adverse environmental impacts associated with tourism activities at tourist sites (Chan, 2013). Conserving the environment is of critical importance given the high susceptibility of the environment to tourism-induced pressure. Ecolabels such as Blue Flag, Green Range, and ISO 14001 have been found to contribute towards sustainability. Worldwide, certification systems are needed to get rid of the damage caused by mass tourism on natural resources (Ozturk & Aydin, 2021). The booming digitisation and virtualisation trends through digital representation is increasingly becoming the main information source of an urbanised and tech-dependent global tourist consumer (Mitova et al., 2021). The next section provides the methods employed in the study.

Materials and Methods

Study Sites and Context

Cultural and heritage attractions are considered as an important area for future tourism development in Zimbabwe. Heritage tourism is in line with the heritage resource-based economic development underpinning the second Republic in Zimbabwe. Currently, Zimbabwe has five WHS, declared by UNESCO. Victoria Falls and Mana Pools are listed in the UNESCO's world heritage centre website (World Heritage List, 2024). Victoria Falls and Mana Pools are two of the famous yet ecologically fragile tourist sites in Zimbabwe, which depend on the natural environment. The sites were retrieved from the WHS global list and Zimbabwe Tourism Authority (ZTA) national list of WHS' most visited sites, which are ecologically fragile. There are reports suggesting that the two sites under the purview of Zimparks are saddled with challenges. The two sites support a wide array of tourism activities involving nature-based tourism. The sites also face challenges such as inadequate resources for maintenance, increased pressure from tourism and urban development, political pressure for the sites to benefit local communities, balancing site preservation and tourism activities, and environmental concerns in terms of degradation.

Victoria Falls is located in Matabeleland North Province and Mana Pools is located in Mashonaland West Province. Both sites are located along the mighty Zambezi River, and support non-consumptive tourism. Mana Pools is at extremely high risk for wildlife poaching and habitat encroachment. Notable challenges constraining sustainable tourism development at Mana Pools include an unsustainable huge elephant population, climate change issues, habitat loss and massive land degradation. Mana Pools boasts of a variety of wildlife species including elephants, wild dogs, buffaloes, leopards, lions, cheetahs, hyenas and over 350 bird species.

Victoria Falls is shared between the borders of Zimbabwe and Zambia, while Mana Pools National Park lies in the flood plain of Zambezi River, Zimbabwe, as shown in the map of Zimbabwe (Fig. 8.1).

Victoria Falls and Mana Pools were officially recognised as designated world heritage tourist centres in 1989 and 1984, respectively, by the United Nations Educational, Scientific and Cultural Organisation (UNESCO) (UNESCO.org). This was based on their natural beauty and intact wilderness. The Zimbabwe Parks and Wildlife Management Authority manages both the world heritage sites. In terms of size, Victoria Falls is 56,300 hectares (Keep Victoria Falls wild.org) and Mana Pools covers 219,600 hectares (Bird Life International, 2023).

Data Collection

A qualitative research approach was adopted. The chapter is based on data collected from15 interviews with key informants selected using purposive sampling procedures. Gnanapala and Sandaruwani's (2016) semi-structured interviews informed the present study. The target population were 20 managers directly involved in tourism revolving around the two WHS. The study is grounded in interpretative qualitative research methodologies, and uses a case-study design. The composition of participants encompassed tourism experts, safari managers, marketing managers, marketing executives, non-consumptive tourism managers, tour operator managers, lodge managers, campsite managers and hotel managers, among

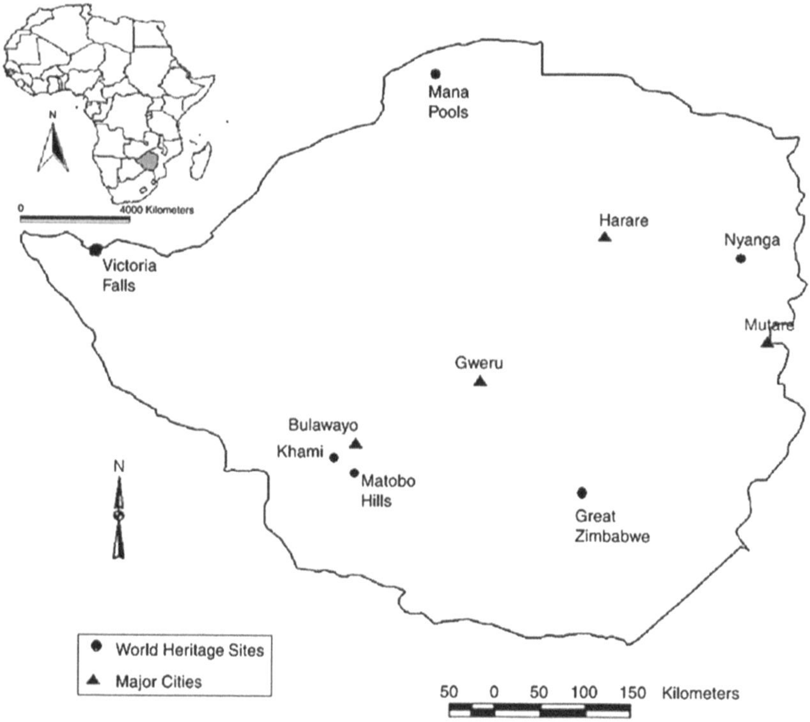

Fig. 8.1 Zimbabwe world heritage sites

others. These specific sectors were deliberately targeted as they reside and operate at the two world heritage sites, and thus had first-hand information of the sustainable marketing practices adopted. To ensure trustworthiness in data collection, only those who benefit directly from tourism activities connected to WHS participated. Additionally, the same question themes were consistently maintained in interview sessions.

The formulation of interview question themes was aligned with research questions. The two study sites are affected by over-tourism, so data are relevant to answer research questions. Selection of participants was through a combination of purposive and snowball sampling procedures. Semi-structured interview sessions lasted 45 minutes, and were conducted on site in the participants' offices. Eight interview sessions were conducted face to face, while seven were telephone interviews. For face-to-face interviews appointments were made at least a week before fieldwork. Telephone interviews were conducted at the convenient time as suggested by interviewees. We resorted to telephone interviews since the targeted interviewees could not grant face-to-face interviews owing to busy work schedules and other commitments. Some of the interviewees' work stations were not easily accessible especially in Mana Pools. We had to reschedule telephone interviews with such targeted participants. In order not to violate ethical protocol, each interviewee was asked to give verbal consent before the recordings of interviews began. The participants were also informed about voluntary participation, and that they could withdraw from the interview, about their right to privacy, and that the data would be used solely for academic purposes. Data were recorded verbatim with the consent of participants using a mobile phone device, and later on transcribed. Interview questions asked encompassed a variety of themes including the specific sustainable marketing practices adopted, significance of the adopted sustainable green marketing practices on customer satisfaction, loyalty, well-being of local communities, impact of sustainable green marketing practices on waste reduction, tourist inflows, product life cycle, resource regeneration and challenges faced in implementing sustainable green marketing strategies, among others. These themes were informed by the study's research questions and objectives, literature review and related past studies. Document analysis was also utilised. Data analysis followed Williams and Moser (2019) thematic content analysis. The next section presents and discusses findings.

Results and Discussion

In this section findings are concurrently presented and discussed in themes.

Sustainable Green Marketing Practices Adopted at World Heritage Sites

Findings showed that a wide array of sustainable green marketing practices have been adopted at Victoria Falls and Mana Pools world heritage sites. These are summarised by the following excerpts:

> We use ecolabels, green certification and ISO certification as part of green marketing strategies. (Zimparks marketing manager, P3)

> Our suite of sustainable marketing practices at Victoria Falls encompass use of digital marketing platforms, eco-tours and Internet or e-marketing. (Marketing executive at Victoria Falls, P10)

> Recycling, zero littering policy, non-consumptive tourism model, use of green energy and polluter pay principle have been infused as sustainable green marketing practices at the two world heritage sites. (Safari company general manger, P12)

The findings reflect emerging marketing trends in the wider global environment in line with the tenets of a circular economy and three pillars of sustainable tourism development. Geremew et al. (2017) underscored the importance of marketing to tourist destination and sites. Worldwide, WHS have adopted a wide array of marketing efforts using digital marketing platforms, printed materials, trade fairs and word-of-mouth advertising, for example, at Amhara. Eco-labelling has been identified as one of the tools for achieving sustainable tourism development in a circular economy. It has become a norm for enterprises to search ways through which sustainable development could be achieved and to prosper in a greener economy by adopting green practices across economic sectors. Sustainable tourism development aims to optimally balance key elements such as economic development, social needs, and biodiversity conservation and ecosystem services. Participants further outlined that Mana Pools and Victoria Falls also use green branding communication strategies whereby they associate the brand with pleasant and environmental imagery of nature, that is, eco-advertising. This is carried out through diverse media outlets to encourage tourists to visit environmentally friendly tourist sites/WHS. According to Rahbar and Wahid (2011) the three elements of eco-labelling are concern for the environment, environmental protection and positive actions of a business on the environment.

Mitova et al. (2021) outline that commercialisation and marketing are significant in making WHS known. To this end, the use of digital marketing at both Victoria Falls and Mana Pools is consistent with what Pineiro-Otero and Martinez-Rokim (2016) refer to as e-marketing. Such digital marketing practices have been hailed for their ability to regulate the number of people visiting the sites, therefore protecting the site in the long term. Mana Pools and Victoria Falls use e-marketing strategies to promote the WHS. According to Zaimes et al. (2022) the use of digitally based marketing strategies has many benefits, such as enabling the sites to gravitate towards a more sustainably inclined tourism model, and such a model aligns with SDGs. Customers are in a position to do virtual tours, which significantly reduces the influx of visitors and physical presence which is stressful to the environment (UN, 2021).

Impacts of Adopted Sustainable Green Marketing Practices on Customer Satisfaction, Loyalty and Well-Being of Local Communities

The study participants perceived the adopted sustainable green marketing practices to have a profound impact on customer satisfaction, loyalty and well-being of local communities. The following narratives summarises this perspective:

> We have enhanced customer loyalty and profit margins. (Non-consumptive tourism manager, P14)
>
> Adoption of zero tolerance to littering has resulted in a clean environment and customer satisfaction. (Marketing executive, P12)
>
> We can charge premium prices leveraging on a clean and quality product offering. (Tour operator manager, P 10)
>
> Local communities' well-being has improved given that they earn income from tourism activities supported by WHS. (Hotel manager, P14)

Customer satisfaction levels are boosted given that number of tourists is managed and kept at the bare minimum, hence no overcrowding. In line with the high yield low volume (HYLV) tourism philosophy, upon which the concept of sustainable tourism development is anchored tourism operators are able to extract more value from very few customers. In the long term this ensures sustainable tourism development as resources are protected. This also aligns with the main pillars of the circular economy which hinges on reuse, resource regeneration and extending product life cycles. This is also in synch with the United Nations 2030 sustainable development goals (SDGs) of engaging in tourism development that support well-being of communities through decent jobs and income. Broadly, a clean environment with no litter enhances tourist experiences, leading to satisfaction. There have been increased positive customer reviews, courtesy of the implementation of sustainable green marketing practices. This enables tourism operators to charge premium prices for a quality product, therefore maximising revenue. This translates to the well-being of local communities. A quality product enables tourism operators to target high net worth customers who spend more and are prepared to pay a premium for a quality product. Customer loyalty results in more business and reduced marketing budgets on the part of tourism operators. Community well-being is enhanced in the sense that more revenue will trickle into the community, and create more sustainable jobs supported by a well-satisfied customer base. Generally, the environment free from pollution, noise, littering and overcrowding enhances the quality of local community life. There is no overstretching of resources at community level, an attribute which mars tourist sites characterised by overcrowding, there is no artificial inflation as visitor numbers are properly managed. In many instances, tourist sites associated with over-tourism (massive influx of tourists) result in local communities' well-being compromised owing to shortage of and competition for resources, a host of social ills, demonstrations, resentment of tourism and everything that it stands for. Adoption of sustainable green marketing practices fosters a harmonious relationship between tourists and host communities, and guard against unequal master-servant relationships. It also ensures that local host communities are not exploited by tourism, resulting in equal relationships. No overexploitation of resources, which allows regeneration and extends the life cycles of products in line with the tourist area life cycle framework. Overall, the finding is in synch with the theories which underpinned the study: circular economy, sustainable tourism development and sustainable green marketing practices as expounded in extant literature.

Impact of Sustainable Green Marketing Practices on Waste Reduction, Product Life Cycles and Resource Regeneration

The adopted sustainable green marketing strategies were commended by most participants for significantly reducing waste disposal at and around the WHS. The following excerpts illustrate this:

> Green marketing strategies have significantly improved the quality of the product much to the satisfaction of our customers. (Safari manager, P13)

> We are getting positive customer reviews particularly around the aspect of environmental friendly tourism orientation. We have earned many accolades from customers because of the way we package our product. (Victoria Falls National Park area manager, P1)

> Few tourists at specific sites allow regeneration of flora. (Zimparks area manager)

> Overall, in my opinion, the adopted marketing practices have extended to life cycles of the heritage sites. (Marketing executive)

Participants outlined that the strict 'polluter pays principle' instilled a sense of responsibility on tourists not to pollute the environment. In addition, the availability of waste disposal bins at strategic points and principle of recycling waste paper greatly improved a clean environment. Most participants were of the view that managing the number of visitors at any given point in time had given the WHS a new lease of life, thus extending the product life cycle. Instead of allowing an uncontrolled influx of tourists, which lead to saturation, coupled with service failures leading to dissatisfied customers who will not return again or recommend the facility to others, the aspect of closing off the facilities at certain times of the year gives sufficient time for the regeneration of vegetation and other related natural resources. This allows a period where the environment is not trampled upon, subjected to less traffic, hence enable it to recover and other ecosystems to reboot/restore. Restricting access in certain areas or spots allow resource rejuvenation. This is in line with the tourist area life cycle, and the concept of sustainable tourism development. It fosters sustainable use of resources, protection of resources, enhances the quality of the tourism product on offer, waste reduction, extends product life cycles, promotes resource regeneration through planned and deliberate breaks or off-season to allow resources to regenerate and give new lease of life for the WHS. These aspects are at the core of a circular economy, sustainable tourism development and sustainable green marketing philosophy.

Increases in socioeconomic activities and tourism infrastructure construction inspired or fuelled by the booming tourism sector are not conducive to the resilience of wildlife habitats at WHS, thereby stressing the ecological system and threatening sustainable ecotourism (Choi et al., 2017). The use and implementation of sustainable green marketing practices at WHS enable stress absorption, fosters regeneration of resources and significantly reduces adverse impacts of tourism activities on tourist attractions. This is in line with the key pillars of the circular economy of recycle, waste reduction and resource regeneration as outlined in extant literature.

Additionally, the use of high-quality eco-educational programme helps tourists become hard eco-tourists. Sustainable green marketing practices ensure that WHS as major tourism resources are protected and preserved, and contribute to local economic vitalisation and support local communities' livelihood. Damage to ecological system adversely influences the social system and threatens the sustainability of ecotourism over the long term (Choi et al., 2014). Sustainable green marketing practices benefit in terms of ecological, economic and socio-cultural aspects, such as benefits contributing to the conservation of wildlife habitats by multiple actors who are involved in and influence tourism activities at WHS. Managing and implementing sustainable green marketing practices ensure that WHS are not exposed to unintended stress which may compromise regeneration of resources and shorten tourist sites' life cycles. Sustainable green marketing practices are predicated on maximising benefits and minimising stress. Negative aspects such as tourist noise, environmental pollution and contaminant accumulation are therefore managed.

Major Challenges Faced by Tourism Operators in Implementing Sustainable Green Marketing Practices at World Heritage Sites

A diverse range of challenges were outlined on this aspect. Participants cited the following as notable challenges facing tourism operators in implementing sustainable green marketing practices: climate change, land degradation, alluvial mining, poaching, limited budgets, political interference, red tape, need to balance resource conservation goals and profit, among others. The study participants outlined a diverse range of factors, which pose a significant threat to the long-term sustainable tourism development at the two WHS. This is supported by the following narratives:

> Alluvial gold panning in the flood plain of the mighty Zambezi River, both in Zambia and Zimbabwe, lately gas mining exploration in Zimbabwe has intensified. (Campsite manager, P7)

> Habitat loss and degradation on both WHS, poaching, pollution, climate change-induced siltation of pools and drying of Falls are significant challenges causing huge concerns on sustainable tourism development. (Marketing manager, P5)

> Cases of increased tourist inflows, over-tourism/tourism, overdevelopment during the peak season are overwhelming the world heritage sites' carrying

capacity. We are seeing potential challenges in the long term provided this trend is maintained. (Marketing executive, P3)

With respect to Mana Pools dam construction activities upstream, destruction of habitat through construction of tourist accommodations such as campsites, veld fire outbreaks, overuse of biological resources, uneven wildlife distribution, climate change issues and poaching remain chronic challenges which threaten long-term sustainable tourism. (Lodge manager, P8)

Most participants were of the view that gas mining exploration along the Zambezi River, reconfiguration of the natural environment by building artificial structures such as sporting facilities and hotels are major challenges faced by the WHS into the foreseeable future. At Victoria Falls participants stated that plans are underway to construct a cricket sporting arena ahead of the 2027 games. Such massive developments tend to significantly alter the natural environment and interfere with the ecosystem. This directly contradicts the tenets of a circular economy, sustainable tourism development and sustainable green marketing practices. It is important to take this into consideration in the marketing of WHS facing such significant challenges. Victoria Falls receives an approximate of over a million tourists per year (Silver Back Gorilla Tours.com). According to Global Conservation (2024), approximately 7000 tourists visit Mana Pools National Park and generate over $2 million of tourism receipts. However, an unsustainable influx of tourists can adversely damage the environment and related ecosystem. Within the context of Victoria Falls, it was highlighted that the increasing number of visitors to Victoria Falls is a source of noise pollution attributed to helicopters, microlight aircraft and boats, and constantly disturbs marine wildlife through riparian activities in the Zambezi River (World Heritage Centre and IUCN (UNESCO World Heritage Centre, 2007). In addition, Mana Pools is also under the threat of gas and oil exploration (Bird Life International, 2023). Pollution of the water sources, loss of critical habitat, habitat alteration, accelerated erosion and increased human disturbance were highlighted as negative impacts to conservation (Bird Life International, 2023). Tourism is characterised as a threat to world heritage sites. For example, there was a strong concern from WHS bodies, pertaining to Great Zimbabwe monuments, whereby an advisory was issued against the idea of putting in place tourism facilities until full excavation was completed. A similar concern was also raised in Peru in which WHS authorities were critical of the rehabilitation project aimed at restoring Chan Archaeological Zone, 'to satisfy the demands of mass tourism'. This is relevant in the context of planned developments at Victoria Falls of sporting facilities, golf courses and additional 5-star hotels. This can be equally interpreted as intention to meet demands of mass tourism. Kilimanjaro National Park in Tanzania is threatened by high tourism intensity, and a similar scenario unfolds at Victoria Falls WHS, whereby issues of littering and facilities maintenance are increasingly becoming a challenge. Increased tourism-related activities potentially lead to over-tourism. Over-tourism naturally results in increases of infrastructural development, which cause significant changes to the environment. At a local level, resources are overstretched and trigger a chain of other challenges. Adverse impacts are inevitable. Ecosystem degradation constrains the environment, altering wildlife habitats, accelerating biodiversity loss as well as interfering with wildlife migration patterns (Stefanica & Butnaru, 2015).

The tragedy of most WHS is the unavailability of proper tourist management strategies to manage the distribution of visitors. WHS in developing countries especially in Africa in general and Zimbabwe in particular are seen as threatened by tourism. In essence, establishing accommodation facilities for tourists leads to widespread changes to the environment. Furthermore, Kaya Kinondo Forest was recommended to be delisted from the WHS list owing to grave concerns around worrisome threats attributed to tourism development, which had cleared the natural forest and replaced it with man-made tourist facilities. Also related to this is the impact of such developments on the area's natural beauty—unsightly features. Mana Pools is located in an ecologically sensitive zone, which is not resilient enough against unrelenting tourism activities. The use of 4-by-4 vehicles in a semi-arid national park adversely affects the environmental ecosystem. The problem is exacerbated by the reality that most developing countries lack some bureaucratic technical capacity to implement sustainable tourism development (Schmutz & Elliott, 2016). The findings are consistent with results in other geographical contexts. For example, similar problems such as over commercialisation of tourism, conflicts of interest and modifications were identified in Sri Lanka's WHS. These have attracted negative publicity about the WHS. Huang et al. (2019) argued that every business entity aims to achieve sustainability; however, challenges such as damaging WHS, commercialisation and other related challenges will always be associated with tourism development at world heritage sites.

Existing Opportunities for Enhancing the Effectiveness of Sustainable Green Marketing Practices

Most participants were optimistic on this aspect. It was stated that tourism operators can leverage on existing legislation to enhance the effectiveness of the adopted sustainable marketing practices. Such legislation includes the Parks and Wildlife Act, Tourism Act and other various acts and statutory instruments, among others. In addition, the inclusion of the two sites on the UNESCO world heritage list presents another opportunity to enhance the effectiveness of the sustainable green marketing practices. Still, the cyclic water flow regime in the Zambezi River can be utilised in the sense that during low seasonal water flow, it can be used to allow resource regeneration.

The packaging of tourism itineraries is also affected. The findings are consistent with earlier results from climate change-related studies in Victoria Falls and other nature-based tourism that climate change is causing physical changes on tourist sites and other recreational facilities (Dube & Nhamo, 2019; Kupika et al., 2018; Zhou, 2023). It becomes urgent and necessary to put in place an intervention plan aimed at restoring the physical appearance of the sites given that a lot is at stake.

Implications on Theory, Policy and Practice

The study has contributed significantly by addressing somewhat limited studies around the specific area of sustainable green marketing practices at WHS in Zimbabwe and in the region. Consequently, the chapter has provided invaluable insights on WHS custodians on specific green marketing practices, which can be implemented to safeguard and protect the sanctity of the sites in the face of growing challenges. These findings can be implemented by other world heritage sites in other southern African countries and beyond which face similar dilemmas. Despite the fact that only two case studies were employed, conclusions can be applied in diverse tourist regions, and relevant in the overall governance of nature-based tourist destinations. The insights of this study can provide valuable guidelines for the effective use of natural resources at WHS.

Conclusion

The study aimed at examining the sustainable green marketing practices at two world heritage sites in Zimbabwe, namely Victoria Falls and Mana Pools. It sought to analyse the impacts of adopted sustainable green marketing practices on customer satisfaction, loyalty and well-being of local communities; examine the significance of sustainable green marketing practices on waste reduction, product life cycles and resource regeneration; determine the major challenges faced by tourism operators in implementing sustainable green marketing practices at World Heritage Sites, and identify existing opportunities for enhancing the effectiveness of sustainable green marketing practices. It also sought to analyse the challenges which threaten long-term sustainable tourism development at the two ecologically fragile world heritage sites. Furthermore, the chapter established that a diverse range of sustainable green marketing practices have been adopted, most notable are the use of ecolabels, green certificate, ISO certification, digital marketing, zero littering policy and adoption of non-consumptive tourism model, among others. Additionally, the study identified over-tourism, climate change issues, gas mining exploration and gold panning, building of dams, sporting facilities and tourist accommodation, habitat loss, and land degradation as major challenges endangering long-term sustainable tourism development at the two world heritage sites. Still, the study found that the adopted sustainable green marketing practices had significantly improved the WHS lifecycles, allowed resource regeneration and reduced waste disposal. Furthermore, the level of customer satisfaction improved significantly, and translated to favourable customer reviews. Local community well-being was also impacted positively.

Study Limitations

This chapter has limitations. It must be acknowledged that ecolabel information from participants was marginal. Secondly, the study exclusively focused on natural world heritage sites excluding man-made world heritage sites. Future studies need to include man-made world heritage sites to broaden our understanding of the green marketing practices adopted to protect the sanctity of the WHS. Finally, a modest literature contribution has been made to the evolving theme of sustainable tourism focusing on world heritage sites, which has been underrepresented in extant literature around WHS in Zimbabwe.

References

Acampora, A., Ali, F., Lucchetti, M. C., Merli, R., & Preziosi, M. (2018). *The impact of green practices on coastal tourism: An empirical investigation on an eco-labelled beach club* (pp. 471–482). Roma Tre University of South.

Adie, B. A. (2017). Franchising our heritage: The UNESCO world heritage brand. *Tourism Management Perspectives, 24*,48–53

Aguiler-Morales, S. Y., Negrete-Cardoso, M., Rosano-Ortega, G., Sanchez-Ruiz, F. J., Sanchez-Baltasar, L. B., Vega-Lebrun, C. A. S., & Chabs-Retchkiman, P. S. (2023). Marketing strategies for waste recycling: a bibliometric analysis towards the circular economy. *Environmental Science and Pollution Research, 30*, 67565–67581.

Alharthey, B. K. (2019). Impact of green marketing practices on consumer purchase intention and buying decision with demographic characteristics as moderator. *International Journal of Advanced and Applied Science, 6*, 62–71.

Amoako, G. K., Dzogbenuku, R. K., Doe, J., & Adjaison, G. K. (2020). Green marketing and the SDGs: emerging market perspective. *Marketing Intelligence and Planning, 40*, 310–327

Asmelash, A. G., & Kumar, S. (2019). *The structural relationship between tourist satisfaction and sustainable heritage tourism development in Tigrai, Ethiopia.*

Atkinson, L., & Rosenthal, S. (2014). Signalling the green sell: the influence of eco-label service, argument specificity and product involvement on customer trust. *Journal of Advertising, 43*, 33–45.

Ban, O., Iacobas, P., & Nedelea, A. M. (2016). Marketing research regarding tourism business readiness for eco-label achievement: Case study: Natura 2000 crisul repede gorge-padurea craiului pass site, Romania. *Ecoforum, 5*(1).

Bird life International Org. (2023). *Oil exploration threatens pristine man pools in Zimbabwe.* Retrieved July 10, 2024, from https://www.birdlife.org/news/2023/06/13/oil-exploration-threatens-pristine-mana-pools-in-zimbabwe/

Brundtland Commission Report. (1987). *Our common future: World commission on environment and development.* Oxford and New York Oxford University Press.

Cavallo, M. (2015). Territorial marketing, green marketing, industrial areas. In M. Caroli, M. Cavallo, & A. Valentino (Eds.), *Eco-industrial parks: A green and place marketing approach* (pp. 49–57). Luiss University Press.

Cellini, R. (2011). Is UNESCO recognition affective in fostering tourism? A comment on Yang, Lin and Han. *Tourism Management, 32*, 452–454.

Chan, E. S. W. (2013). Gap analysis of green hotel market. *International Journal of Contemporary Hospitality Management, 25*(7), 1017–1048.

Choi, E., Song, K., Kim, M., & Lee, J. (2017). Transformational planning for resilient wildlife habitats in ecotourism systems. *Sustainability, 9*, 487. https://doi.org/10.3390/su9040487

Chamberlin, L., & Boks, C. (2018). Marketing approaches for a secular economy: using design frameworks to interpret online communications. *Sustainability, 10*, 2070.

Choi, Y. E., You, S., Ham, E. K., Song, K., & Chon, J. (2014). Sustainable management planning of ecotourism resources using system thinking: A case study of Bamseom in Seoul. *Journal of Tourism Sciences, 38*, 59–79.

Conradin, K., & Wiesmann, U. (2014). Do world natural heritage sites trigger sustainable regional development efforts? *Journal of Protected Mountain Areas Research and Management., 6*, 5–12. https://doi.org/10.1553/ecomont-6-255

Cuccia, T., Guccio, C., & Rizo, I. (2013). *Does UNESCO inscription affect the performance of tourism destination? A regional perspective*. ACEI Working Paper Series AWP-04-.

D'Souza, C., Taghian, M., & Lamb, P. (2006). An empirical study on the influence of environmental labels on consumers. *Corporate Communication International Journal, 11*(2), 162–173.

Delgado-Torres, A., & Polomeque, F. L. (2012). The growth and spread of the concept of sustainable tourism: The contribution of institutional indicators to tourism policy. *Tourism Management Perspectives, 4*, 1–14.

Dodds, R., & Jopp, M. (2001). Promoting urban green tourism: The development of the other map of Toronto. *Journal of Vacation Marketing, 7*(3), 261–267.

Dube, K., & Nhamo, G. (2019). Climate change and potential impacts on tourism: Evidence from the Zimbabwean side of the Victoria Falls. *Environmental, Development and Sustainability, 21*, 2025–2041.

Dube, K., & Nhamo, G. (2021). Greenhouse gas emissions and sustainability in Victoria Falls: Focus on hotels, tour operators and related attractions. *African Geographical Review, 40*(2), 125–140.

Dunk, R. M., Gillespie, S. A., & MacLeod, D. (2016). Participation and retention in a green tourism certification scheme. *Journal of Sustainable Tourism, 24*(12), 1585–1603.

Elliott, M. A., & Schmutz, V. (2012). World heritage: Constructing a universal cultural order. *Poetics, 40*, 256–277.

Espron, M., Gyurus, E., & Stoeckl, N. (2014). Does eco-certification deliver benefits? An empirical investigation of visitors' perceptions of the importance of eco certificates' attributes and of operations' performance. *Journal of Sustainable Tourism, 22*(1), 148–169.

Fonsera, L. M., Domingues, J. P., & Dima, A. M. (2020). Mapping the SDGs relationships. *Sustainability, 12*(8).

Font, X., Elgammal, I., & Lamond, I. (2017). Green hushing: The deliberate under communicating of sustainability practices by tourism business. *Journal of Sustainable Tourism, 25*(7), 1007–1023. https://doi.org/10.1080/09669582.2016.1158829

Hassan, S. (2000). Determinants of market competitiveness in an environmentally sustainable tourism industry. *Journal of Travel Research, 38*(3), 239–245.

Geremew, Y. M., Alemayehu, H. B., & Meskede, A. T. (2017). Tourism marketing challenges and new tourism product development potentials: Practices from Amhara natural regional state world heritage sites. *Journal of Hospitality and Management Tourism, 8*(1), 1–13.

Gidey, Y., & Sharma, K. (2017). Tourists satisfaction in tourist destination- a study of Tigray-Ethiopia. *International Journal of Research in Finance and Marketing, 4*, 138–174.

Global Conservation. (2024). Retrieved July 10, 2024, from https://globalconservation.org/projects/mana-pools-world-heritage-site-zimbabwe

Gnanapala, W. K. A., & Sandaruwani, J. A. R. C. (2016). Impacts of tourism development in cultural and heritage sites; an empirical investigation. *International Journal of Economics and Business Administration, 2*(6), 68–78.

Gonzales, G. (1996). *Sustainable tourism: the new tourism paradigm.* Retrieved from https://www.gonzales.com.sg/sustour.html

Gryshchenko, O., Babenko, V., Bilovodska, O., Voronkova, T., Ponomarenko, I., & Shatskaya, Z. (2022). Green tourism business as marketing perspective in environmental management. *Global Journal of Environmental Science Management, 8*(1), 117–132.

Handriana, T., & Ambara, R. (2016). Responsible environmental behaviour intention of travellers on ecotourism sites. *Tourism and Hospitality Management, 22*(2), 135–150.

He, G., Zhao, X., & Yu, M. (2021). Exploring the multiple disturbances of karst landscape in Guilin world heritage site, China. *Catena*. https://doi.org/10.1016/j.catena2021.105349

Huang, R. Y., Chang, W. J., Chung, Y. C., Yin, Y. S., & Yen, J. (2019). A literature review of sustainable tourism, 1990–2016. *International Journal of Tourism and Hospitality Reviews, 6*(1), 20–49.

ISO. (2000). ISO 14020:2000. *Environmental labels and declarations- general principles.* International Standards Organisation.

IUCN. (2017). Conservation outlook assessment: Wulingyuan scenic and historic area. http://worldheritageoutlook.iucn.org/explore-sites/wdpaid/67731

Jayakumar, C. (2017). *Green marketing and sustainability.*

Joshi, Y., & Rahman, Z. (2015). Factors affecting green purchase behaviour and future research directions. *International Strategic Management Review, 3*, 128–143.

Jurado-Rivas, C., & Sanchez-Rivero, M. (2019). Willingness to pay for more sustainable tourism destination in world heritage cities: The case of Cacerav. *Sustainability, 11*, 5880.

Juvan, E., & Dolnicar, S. (2014). The attitude-behaviour gap in sustainable tourism. *Annals of Tourism Research, 48*, 76–95.

Keep Victoria Falls Wildv. (2024). *A brief history of the Victoria Falls National Park*. Retrieved July 10, 2024, from https://keepvictoriafallswild.com/victoria-falls/newdevelopments/vfnp.html

Kenterelidou, C., & Galatsopoulou, F. (2021). Sustainable biocultural heritage management and conservation: The case of digital narrative for UNESCO marine world heritage of outstanding universal value. *Sustainability, 13*(3), 1449. https://doi.org/10.3390/su13031449

Kim, H., Stepchenkova, S., & Yilmaz, S. (2019). Destination extension: A faster route to fame for the emerging destination brands? *Journal of Travel Research, 58*(3), 440–458.

King, L. M., & Halpenny, E. A. (2014). Communicating the world heritage brand: Visitor awareness of UNESCO's world heritage symbol and the implications for sites, stakeholders and sustainable management. *Journal of Sustainable Tourism, 22*, 768–786.

Kinoti, M. (2011). Green marketing intervention strategies and sustainable development. *International Journal of Business and Social Science, 2*(23), 27–47.

Kirchherr, J., Reike, D., & Hekkert, M. (2017). Conceptualising the circular economy: an analysis of 114 definitions. *Resources, Conservation and Recycling, 127*, 221–232.

Kumar, V., Rahman, Z., Kazmi, A. A., & Goyal, P. (2012). Evolution of sustainability as marketing strategy: Beginning of new era. *Procedia-Social and Behavioral Sciences, 37*, 482–489.

Kupika, O. L., Gandiwa, E., Kativu, S., & Nhamo, G. (2018). *Impacts of climate change and climate variability on wildlife resources in southern Africa: Experience from selected protected areas in Zimbabwe*.

Lai, P. H., & Shafer, S. (2005). Marketing ecotourism through the internet. An evaluation of selected eco-lodges in Latin America and the Caribbean. *Journal of Ecotourism, 4*, 143–160.

Lee, S. M., Honda, H. C., Ren, G., & Lo, Y. C. (2016). The implementation of green tourism and hospitality. *Journal of Tourism and Hospitality, 5*(4), 1–5.

Liao, Y. K., Wu, W. Y., & Pham, T. T. (2020). Examining the moderating effects of green marketing and green psychological benefits on customer's green attitude, value and purchase intention. *Sustainability, 12*, 7461.

Madhushika, G. H. P. H., & Sivesan, S. (2022). Developing sustainable marketing framework for world heritage sites in Sri Lanka. *Journal of Tourism economics and applied research, 6*(II), 71–83.

Margottini, & Vilimek. (2024). The ICL network on landslide and cultural and natural heritage (LACUNHEN). *Landslides, 11*(5), 933–938.

Merrrilees, B. (2016). Interactive brand experience pathways to customer-brand engagement and value co-creation. *Journal of Product and Brand Management, 25*(5), 402–408.

Mitova, R., Borisova, B., & Koulov, B. (2021). Digital marketing of Bulgarian national heritage for tourism and recreation. *Sustainability, 13*, 13071.

Mkono, M. (2019). The future of tourism in Africa: Optimism in a changing environment. In M. Mkono (Ed.), *Positive tourism in Africa* (pp. 245–251). Routledge.

Muposhi, A., Dube, K., & Nyakudya, C. (2024). Disengaging with green hotels: Behavioural reasoning, willingness to pay and patronage intention: A developing country perspective. *International Journal of Tourism Policy, 14*(5), 493–507.

Negrete-Cardoso, M., Rosano-Ortega, G., Alvarez-Aros, E. L., Tavera-Cortes, M. E., Vega-Lebrum, C. A., & Sanchez-Ruiz, F. J. (2022). Circular economy strategy and waste management: a bibliometric analysis in its contribution to sustainable development towards a post COVID-19 era. *Environmental Science and Pollution Research.*

Nekmahmud, Md., & Fekete-Farkas, M. (2021). Green marketing, investment and sustainable development for green tourism. In A. Hassan (Ed.), *Tourism in Bangladesh: Investment and development perspectives* (pp. 339–362). Springer Nature.

Njerekai, C. (2019). Hotel characteristics and the adoption of demand oriented hotel green practices in Zimbabwe: A regression analysis. *African Journal of Hospitality, Tourism and Leisure, 8*(2), 1–11.

Nyahunzvi, D. K. (2014). Greening Gweru urban tourist lodges: Low levels of ecoliteracy, green awareness and actions. *African Journal of Hospitality, Tourism and Leisure.*

Ottman, J. A., Stafford, E. R., & Hartman, C. L. (2006). Avoiding green marketing myopia: Ways to improve consumer appeal for environmentally preferable products. *Service and Policy for Sustainable Development, 48*(5), 22–36.

Ozturk, A., & Aydin, I. Z. (2021). Assessments of certification systems in forestry and sustainable ecotourism. *Socrates Journal of Interdisciplinary Social Studies, 7*(9), 1–13.

Pardana, D., Abdullah, R., Mahmuda, D., Malik, E., ... & Hamid, R. S. (2019). Attitude analysis in the theory of planned behaviour: green marketing against the intention to buy environmentally friendly products. *IOP Conference Series Earth and Environmental Science, 343*, 012128.

Pineiro-Otero, T., & Martinez-Rokim, X. (2016). *Understanding digital marketing-business and actions* (pp. 37–74). MBA Springer.

Polonsky, M. J. (1994). An introduction to green marketing. *Electronic Green Journal, 2.*

Poranonond, P., & King, V. T. (2014). *Rethinking Asian tourism: Culture, encounters and local response.* Cambridge Scholars Publishing.

Poria, Y., Reichel, A., & Cohen, R. (2013). Tourists' perceptions of world heritage sites and its designation. *Tourism Management, 35*, 272–274.

Proto, M., Malandrino, O., & Supino, S. (2007). Eco-labels: A sustainability performance in benchmarking? *Management of Environmental Quality International Journal, 18*(6), 669–683.

Rahbar, E., & Wahid, W. A. (2011). Investigation of green marketing tools' effect on customers' purchase behaviour. *Business Strategy Series, 12*(2), 73–83.

Ruskule, A., Klepers, A., & Veidemane, K. (2018). *Mapping and assessment of cultural ecosystem services of Latvian coastal areas.* Baltic Environmental forum, Riga, Latvia. University of Latvia.

Ryan, J., & Silvanto, S. (2014). A study of the key strategic drivers of the use of the world heritage site designation as a destination brand. *Journal of Travel Tourism Marketing, 31*, 327–343.

Sadiq, M., & Adil, M. (2020). Ecotourism related search for information over the internet: A technology acceptance model perspective. *Journal of Ecotourism, 20*, 70–88.

Samuels, K. L., & Platts, E. J. (2020). An eco-label for the world heritage brand? Developing a climate communication recognition scheme for heritage sites. *Climate, 8*(38).

Sangpikul, A. (2010). Marketing ecotourism through the internet: A case study of ecotourism business in Thailand. *International Journal of Hospitality and Tourism Administration, 11*, 107–137.

Schmutz, V., & Elliott, M. A. (2016). Tourism and sustainability in the evaluation of world heritage sites, 1980-2010. *Sustainability, 8*, 261. https://doi.org/10.3390/su8030261

Schmutz, V., & Elliott, M. A. (2018). World heritage and the scientific conservation of universal value. *International Journal of Comparative Sociology.*

Sedky, D., & AbdelRaheem, M. A. (2021). Studying green marketing in emerging economies. *Business Strategy and Development, 5*, 59–68.

Shandra, J. M., Esparza, L. E., & London, B. (2012). Non-governmental organisations, democracy and deforestation: A cross-national analysis. *Society and Natural Resources, 25*, 251–269.

Shereni, N. C., & Rogerson, C. M. (2022). Sustainable practices and benefits in the hospitality sector of Zimbabwe. *Geojournal of Tourism and Geosites, 43*(3), 1030–1037.

Shereni, N. C., & Saarinen, R. (2020). Community perceptions on the benefits and challenges of community-based natural resources management in Zimbabwe. *Development Southern Africa*, 1–17.

Shi, H., Shi, T., Han, F., Liu, Q., & Zhao, H. (2019). Conservation of natural heritage sites outstanding universal value via multiple techniques-Bogda Xinjiang Tishan. *Sustainability, 11*(21), 5953.

Silver Back Gorilla Tours.com. (Year). *Facts about Victoria falls in Zimbabwe and Zambia.* https://www.silverbackgorillatours.com/tourist-attractions/victoria-falls-in-zimbabwe-and-zambia

Siyabona Africa. (2024). *World heritage sites in Zimbabwe*. https://www.siyabona.com/world-heritage-sites-zimbabwe.html

Stefanica, M., & Butnaru, G. I. (2015). Research on tourists' perception of the relationship between tourism and environment. *Proceedia Economics and Finance, 20*, 595–600.

Su, Y. W., & Lin, H. L. (2014). Analysis of international tourist arrivals worldwide: The role of world heritage sites. *Tourism Management, 40*, 46–58.

UN. (2021). *Sustainable development –The 17 goals*. United Nations. https://sdgs.un.org/goals

UNESCO. (2021). UNESCO World Heritage centre- World Heritage List.

UNESCO World Heritage Centre. (2007). *Mosi-oa-Tunya/Victoria falls*. Retrieved July 10, 2024, from https://whc.unesco.org/en/soc/967/

UNWTO & UNDP. (2018). *Tourism and the sustainable development goals-Journey to 2030*. https://www.undp.org/sites/g/files/zskgke326/files/publications/UNWTO_UNDP_Tourism%20and%20the%20SDGs.pdf

Utami, A. N., Iqbal, A., & Rahab. (2022). Green marketing design strategy for Limpakuwus pine Forest natural tourism, Banyumas city, Indonesia. *Journal of Economics and Sustainability, 4*(1), 18–32.

Wang, Z. G., & Yuan, B. C. (2020). Harmonising the branding strategy of world natural heritage in China: Visitors' awareness of the multiple brands of Walingyuan. *Zhangjiajie Geoheritage, 12*, 1–11.

Williams, M., & Moser, T. (2019). The art of coding and thematic exploration in qualitative research. *International Management Review, 15*, 45.

World Heritage List. (2024). Available online. Retrieved August 2, 2024, from https://whl.unesco.org/en/list/

Wuepper, D., & Patry, M. (2017). The world heritage list, which sites promotes the brand? A big data spatial econometrics approach. *Journal of Cultural Economy, 41*, 1–21.

Yang, C. H., & Lin, H. L. (2010). Analysis of international tourist arrivals in China: The role of world heritage sites. *Tourism Management, 31*, 827–837.

Yilmaz, Y., Unguren, E., & Kacmaz, Y. Y. (2019). Determination of managers' attitudes towards eco-labelling applied in the context of sustainable tourism and evaluation of the effects of eco-labelling on accommodation enterprises. *Sustainability, 11*(5069), 1–27.

Yttredal, E. R., & Homlong, N. (2019). Forbruk blant. Besokende til Geirangeromradet. www.sustrans.no

Zaimes, G. N., Iakovoglou, V., Maclaren, F. T., & Manchanda, P. (2022). Adopting digital tools and technology to evolve sustainable tourism at world heritage sites: Case studies from India and Greece. In M. T. Albert et al. (Eds.), *50 years world heritage Concention: Shared responsibilities-conflict and reconciliation. Heritage Studies*. https://doi.org/10.1007/978-3-031-05660-4_28

Zengeni, N., Zengeni, D. M. F., & Muzambi, S. (2013). Hoteliers' perceptions of the impacts of green tourism on hotel operating costs in Zimbabwe: The case of selected Harare hotels. *Australian Journal of Business Management Research, 2*(11), 64–73.

Zhang, S., Xiong, K., Fei, G., Zhang, H., & Chen, Y. (2023). Aesthetic value protection and tourism development of the world natural heritage sites: A literature review and implication for the world heritage karst sites. *Heritage Science, 11*, 30. https://doi.org/10.1186/s40494-023-0087-0

Zheng, D., Linag, Z., & Ritchie, B. W. (2020). Residents' social dilemma in sustainable heritage tourism: The role of social emotion, efficacy, beliefs and temporal concerns. *Journal of Sustainable Tourism, 28*(11), 1782–1804.

Zhou, Z. (2018). Tourism sector: A bright light in Zimbabwe's depressed economic environment. *African Journal of Hospitality, Tourism and Leisure, 7*(1), 1–19.

Zhou, Z. (2023). Climate change and risks: Insights for tourism development in Victoria Falls, Zimbabwe. *SN Social Sciences, 3*, 159.

Zhou, Z., & Makwara, T. (2024). Challenges facing community-based tourism resuscitation at Dzimbabwe cultural tourism Centre, Zimbabwe: Grass-root perspectives. *SN Social Sciences, 4*, 149.

Zimbabwe world heritage sites. https://link.springer.com/referenceworkentry/10.1007/978-1-4419-0465-2_1963/figures/2614

CHAPTER 9

Advancing Sustainable Green Marketing Agenda in Africa Through Green Supply Chain Management

David Chikwere, *Shakerod Munuhwa*, *and Munyaradzi Chibaro*

Introduction

Globally, sustainable environmental practices have gained traction informed by the sustainable development goals (SDGs) (Agyabeng-Mensah et al., 2020). This has been necessitated by environmental degradation which contributed to climate change and global warming (Sarkodie,

D. Chikwere (✉)
Leeds Business School, Leeds Beckett University, Leeds, UK
e-mail: D.chikwere@leedsbeckett.ac.uk

S. Munuhwa
Department of Management, Glasgow School for Business and Society, Glasgow Caledonian University, Glasgow, UK

M. Chibaro
University of Botswana, Gaborone, Botswana

A. Muposhi, B. Nyagadza (eds.), *Sustainable Green Marketing Strategies for a Circular Economy*, Palgrave Studies of Marketing in Emerging Economies,
https://doi.org/10.1007/978-3-031-81627-7_9

2018; Al-Ghussain, 2019). Consequently, there has been a sharp increase in the number of businesses creating eco-friendly systems, and products for their customers (Yildiz Çankaya & Sezen, 2019).

Green supply chain management (GSCM) has emerged as a vital strategy for promoting sustainability in business operations (Maditati et al., 2018; Namagembe et al., 2019). The goal of GSCM is to incorporate environmental considerations into all aspects of supply chain management, including product design, procurement of raw materials, production procedures, final product delivery, and end-of-life management (Seman et al., 2019; Wang et al., 2020). Green marketing which is the application of sustainability measures for environmental safety and organisational performance in marketing campaigns can be enhanced by GSCM (Chen et al., 2020; Hinson et al., 2021).

The Africa region like other regions in the Global South is facing a twin problem of population growth and depleting resources (Hinson et al., 2021). Therefore, GSCM is critical in advancing green marketing and sustainable development in Africa, where materialism and rapid industrialisation are making environmental challenges increasingly pressing (Abdel-Baset et al., 2019; Namagembe et al., 2019). The African continent's escalating environmental problems are pollution, deforestation, and resource depletion (Sarkodie, 2018). To address the foregoing challenges, more sustainable economic methods should be implemented to achieve the United Nations' 2030 agenda for Sustainable Development and Africa's developmental blueprint, Agenda 2063.

Despite the critical role of GSCM in fostering green marketing, there is a lack of studies that comprehensively address the interaction between these two elements in an African context. The studies which focused on green marketing treated it as an independent domain, with little attention on the function of GSCM in supporting it. This gap in the literature suggests the need for greater exploration into how GSCM may be utilised to fulfil green marketing agenda in Africa. This chapter addresses the literature gap by providing a theoretical foundation for understanding the integration of GSCM into business practices in Africa to foster green marketing practise. This chapter addresses the following objectives:

- Examine the relationship between GSCM and green marketing
- Explore the role of GSCM in enhancing green marketing agenda in Africa
- To recommend best practices of integrating GSCM and green marketing

By exploring the linkages between green marketing and GSCM, African companies stand to benefit from the growing demand of eco-friendly goods and operating efficiencies at the same time reducing negative environmental externalities (Jia et al., 2018; Namagembe et al., 2019).

Literature Review

This chapter is supported by the institutional theory and sustainability theory. The institutional theory is premised on the assumption that institutions are the primary focus of analysis for both organisational behaviour and design (Lammers et al., 2014; Huq & Stevenson, 2020). This viewpoint holds that organisations are merely expressions of larger institutional structures (Lammers et al., 2014). Organisational structures, practices, and design aspects are shaped and directed by institutions which are defined as deeply rooted beliefs, regulations, and standards (Huq & Stevenson, 2020). It is believed that following formalised standards can help organisations become more legitimate, less ambiguous, and more transparent in its decisions which fosters sustainability (Kauppi & Hannibal, 2017; Berthod et al., 2018).

The use of sustainable GSCM and green marketing strategies in Africa can be better understood by looking at the theory of institutions. Organisations in Africa are influenced by institutional decisions from industry norms, regulatory entities, and cultural expectations about an ecologically friendly environment with economic factors (Kauppi & Hannibal, 2017). As institutional theory posits that institutions adhere to external regulations to thrive in their own institutional environments (Peters, 2022), it is critical that governments and organisations in Africa align their laws with sustainability. However, the institutional theory may require some adaptation to meet sustainability situations of organisations unless it is complemented by another theory, such as the sustainability theory. The sustainability theory which has its roots in the 1980s is defined as a three-dimensional balanced state that integrates social, economic, and environmental requirements and their respective saliences (Caradonna, 2022). Real sustainability can only be attained when these three pillars are in harmony and balance with one another (Waring, 2016; Poveda, 2017). It draws attention to how these dimensions are interrelated and emphasises the necessity of all-encompassing methods that address sustainability from all angles at once (Barbier & Burgess, 2017; Poveda, 2017).

De Faria (2014) argues that sustainability should be viewed as a process that values its producers in addition to being a means of creating objective and subjective material conditions for social existence, challenging the conventional notion of sustainability. Starik and Kanashiro (2013) posit that the creation of sustainability management theory is required to address the unique opportunities and issues related to the shift to sustainability. Baeza (2018) takes a comprehensive approach to sustainability by utilising fuzzy logic and set theory which underscores the complexity of sustainability and the necessity for critical thinking and a systemic approach to improve sustainability practice at the individual, organisational, and societal levels. When combined, institutional and sustainability theories offer a strong foundation for comprehending how GSCM can support environmentally friendly marketing strategies that are sustainable. These theories provide a platform on which organisations can enhance their ability to meet and surpass sustainability expectations by utilising green skills. These skills include waste reduction, eco-friendly material utilisation, and resource optimisation (Kauppi & Hannibal, 2017). This strategy encourages a more sustainable future by supporting environmental sustainability as well as the social and economic well-being of local communities.

Green Supply Chain Management Practices

Green supply chains combine environmentally friendly concepts with supply chain management to enhance environmental sustainability through a variety of green practices, such as green manufacturing processes, green end-of-life management for products, green distribution and warehousing, green transportation using biofuels, and green purchasing (Khan et al., 2017). Reduction of greenhouse gas emissions is the major aim of GSCM. However, there are other real benefits that an organisation can experience, such as increased asset efficiency, decreased waste production, increased innovation, lower production costs, reused raw materials, higher profitability, and a sense of added value to the customer base (Cousins et al., 2019). This is accomplished by increased transparency, integration of supply chain systems and procedures, and increased cooperation between all parties and their upstream and downstream supply chain partners (Yildiz Çankaya & Sezen, 2019; Agyabeng-Mensah et al., 2020). Green purchasing, green manufacturing, green distribution, and green reverse logistics are identified as the major forms of green supply chain management practices (Agyabeng-Mensah et al., 2020).

Green purchasing pertains to the acquisition of materials and components that possess eco-friendly attributes including recyclable nature, non-toxicity, and reusability. It is divided into five primary categories which are supply chain management, ecological management, environmental management, environmental authentication, and design operation management (Yang et al., 2010). Achieving both financial and environmental performance is possible when these activities are strictly adhered to (Agyabeng-Mensah et al., 2020; Hashmi, 2023). To pursue reduction in waste and contamination of the air and water, businesses and supply chains can effectively use green purchasing practices (Yildiz Çankaya & Sezen, 2019; Cousins, et al., 2019). For example, 100% of virgin cotton in Patagonia's clothes is responsibly sourced (Patagonia, 2024).

Green manufacturing, also known as green production, is a commitment to use production systems and facilities which have less negative impact on the environment (Bendig et al., 2023). This is achieved through the revitalisation of production methods using green technology and the formation of ecologically sustainable businesses in the manufacturing sector (Borges et al., 2021). For example, Patagonia realised a 100% electricity for its operations in the United States in 2020 from renewable sources (Patagonia, 2021). As a result, employees consume fewer natural resources by using energy from renewable sources, lessen waste and pollution, recycle and reuse materials, and decrease emissions during operations.

It is a concept also known as sustainable distribution which emphasises on efficiency to prevent energy wastage and value addition at each stage across the supply chain. Green distribution seeks to integrate environmental responsibility practices into an organisation's supply chain (Yildiz Çankaya & Sezen, 2019; Agyabeng-Mensah & Tang, 2021). Therefore, distribution routes and transportation routes used within a production cycle should be sustainable given that they have a direct impact on a product's ecological footprint. The distribution system ought to be set up in a way that conserves resources and the environment (Mwaura et al., 2016). Organisations can accomplish sustainability by transparent systems and utilising local raw materials to reduce fossil fuel use and greenhouse gas emissions.

Green reverse logistics seeks to realise value from a product at the end of its economic useful life through recycling and reproduction (Agyabeng-Mensah & Tang, 2021). The greening of reverse logistics assists organisations to achieve sustainable operations (Ali et al., 2020; Hashmi, 2023). This can be done through recovering useful resources from waste for reprocessing, reuse, recycling, or refurbishment thereby reducing amount of garbage, disposal cost, and environmental pollution (Jalil, 2017; Sharma et al., 2021). Organisations can devise strategies for green reverse logistics by changing their product design to foster recyclability, promotion of green packaging to allow reuse, and seamless communication for traceability across the supply chain (Chen et al., 2019).

Green Marketing Concept

Green marketing, also known as responsible marketing, is the promotion of products with eco-friendly qualities. It applies sustainability concepts to a range of marketing activities, such as modifying products, altering the production process, and using sustainable packaging (Dangelico & Vocalelli, 2017; Hinson et al., 2021). Furthermore, green marketing improves a company's financial, environmental, and competitive conditions while enhancing its reputation and brand (Pahlevi & Suhartanto, 2020). For instance, 64% of Singaporean consumers in June 2020 perceive that a brand's environmental responsibility and sustainability were very relevant when making their brand choice (Statista, 2020).

Green marketing, which focuses on preserving resources, supporting communities, and protecting the environment, has emerged as a response to today's environmental challenges (Dangelico & Vocalelli, 2017; Han et al., 2022). There are various ways organisations can employ to build eco-friendly and sustainable operations which enhance a sustainable brand (Pang & Zhang, 2019; Nguyen et al., 2020). By developing products and services through sustainable practices and eco-friendly components, green marketing seeks to address the growing demand for environmentally responsible solutions.

Materials and Method

The Preferred Reporting Items for Systematic Review and Meta-Analysis (PRISMA) technique (Moher et al., 2015) was used to perform a thorough literature assessment on the role of GSCM in fostering green marketing agenda in Africa. To conduct the study, the selected keywords were validated by peers and field specialists. This was followed by searching in the primary sources including the most cited journals on GSCM and green marketing from Scopus, Web of Science, and Google Scholar databases. The chosen publications were those that were published in peer reviewed journals on the topic between 2017 and 2024. There were 421 articles found in the bibliographical search which underwent several stages of filtering. Finally, 20 papers that were deemed most pertinent were selected for assessment.

Research Question Formulation

The Population or issue, Intervention, Comparator, Outcome (PICO) framework was applied to precisely describe the research topic and generate search strings that are compatible with scientific databases (Eriksen & Frandsen, 2018). The PICO framework was critical to systematically evaluate the appropriateness and value of an intervention within a specific population which assisted in organising and refining the research question (Mancin et al., 2024). This approach ensured a comprehensive and precise analysis of the topic.

This methodological tool, which is vital for developing structured research questions for systematic reviews, covers the following important areas: Population refers to the target group or topics, which are organisations in the African region involved in manufacturing, agriculture, retail, or any organisation whose operations have potential to contaminate the environment. A comparator helps to specify a control reference against which the intervention can be evaluated, while an intervention describes the set of actions that will be investigated with the intention of precisely identifying its impact which is the adoption of GSCM. An outcome displays the anticipated measurements or results that the intervention is anticipated to yield which is the green marketing agenda (Methley et al., 2014; Pollock & Berge, 2018).

The Search Strategy

The PRISMA (Preferred Reporting Items for Systematic Reviews and Meta-Analysis) approach was used to carry out the search strategy to ensure accurate and transparent systematic review execution and reporting (Page et al., 2021). The publications were chosen and evaluated according to how well they addressed the goals of the study, looking at concepts and themes associated with the issues that were noted. Recognising current GSCM practices and guidelines was crucial at this point. Furthermore, the creation of research questions using the PICO framework served as the basis for the focused keywords employed or search phrases that are classified in Fig. 9.1 below.

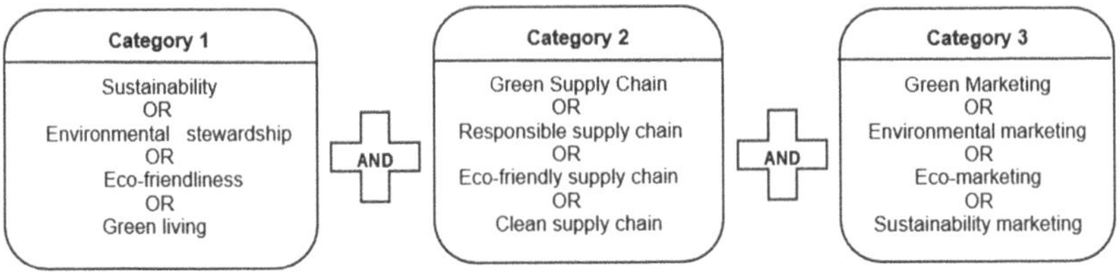

Fig. 9.1 Keywords employed in the literature search

Inclusion and Exclusion Criteria

When determining the scope of the review, it was imperative to specify precise inclusion and exclusion criteria (Patino & Ferreira, 2018; Mancin et al., 2024). The targeted primary studies included papers referring to empirical research that were in line with the study's scope and systematic reviews. This technique ensured that only pertinent, high-quality sources were used which minimised the influence of records that did not fulfil strict quality standards. The inclusion criteria encompassed additional variables such as the type of study, sample size, length of investigation, and analysis of the origin. Studies on the adoption of GSCM by African organisations and articles with possible solutions for GSCM implementation were additional inclusion criteria. Only articles published in English language which is most used by the GSCM research community were taken into consideration to ensure data accessibility and comprehensibility. As depicted on the PRISMA flowchart, publications that have no connection to GSCM or that did not concentrate on Africa were excluded. Publications outside the time range of the systematic literature review and non-academic or non-peer reviewed articles were also excluded. Publications that do not address the GSCM concerns, such as those which only address infrastructure, government, finance, or technology, were subject to further exclusion criteria.

The PRISMA Flowchart

The study adopted the PRISMA framework which was developed by Moher et al. (2015) which is depicted in a flowchart in Fig. 9.2. The drafting process was used to extract the most relevant articles on the adoption of GSCM in Africa. A thorough search of three databases (Scopus, Web of Science, and Google Scholar) produced 421 results as shown in Fig. 9.2.

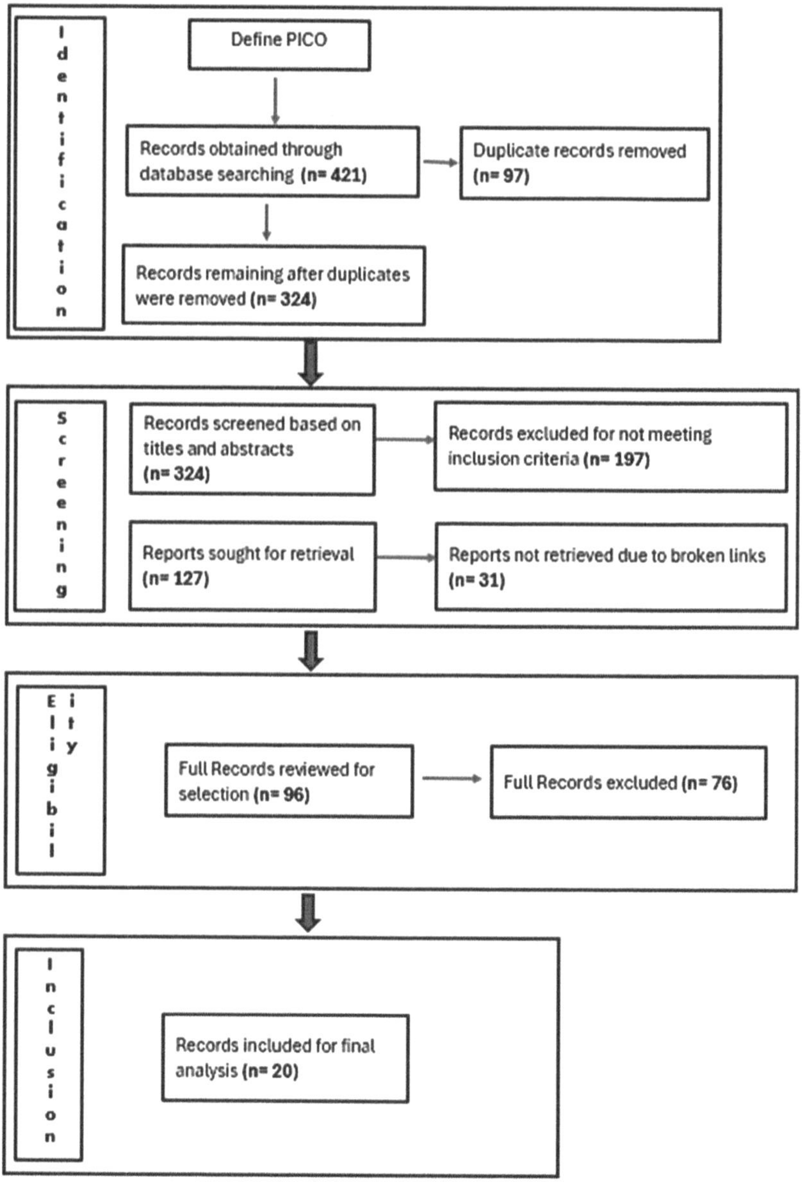

Fig. 9.2 Preferred reporting items for systematic reviews and meta-analysis

There were 324 distinct records left for screening based on their titles and abstracts after 97 duplicate records were removed. Some further 197 of the records were excluded because they did not fit the inclusion criteria as they had no connection to GSCM (53 reports), published outside the considered period (48 reports), missing journals (29 reports), non- peer reviewed sources (18 reports), and not covering issues related to Africa (49 reports). This left 127 of the records to be sought for retrieval. During the retrieval process, some 31 reports could not be accessed because of broken links such as paywalls, subscription restrictions, not peer reviewed, and not being published in English language. Following a thorough evaluation of 96 full reports, the eligibility assessment phase resulted in the exclusion of 76 publications for various reasons, including inadequate data (21 reports), low quality (27 reports), and mismatches with the research objectives (28 reports). In the end, 20 studies were considered for the final analysis which provided a strong basis for comprehending how GSCM advances green marketing agenda in Africa.

Assessment of Quality and Bias

For every publication, two reviewers evaluated it to determine the methodological quality and bias risk. A third review author was included to resolve any disagreements that arose during this phase of the systematic review process. During the database search, at least two researchers searched the database to ensure objectivity and reduce the possibility of bias (Ouzzani et al., 2016; Mancin et al., 2024). A comprehensive evaluation of the selected literature was conducted utilising customised and validated checklists included in the systematic review. A wide range of sources was incorporated to help ensure a balanced understanding of the obstacles to GSCM in Africa and to prevent any potential biases.

Studies Included

Ghana had the highest number of publications with 35% of the total (7 out of 20), followed by South Africa, Nigeria, and Ethiopia which contributed 20%, 15%, and 10%, respectively. The figures point to a leading role and strong focus by these countries in GSCM research area. The other countries that complete the distribution list include Kenya, Tanzania, Uganda, and Zimbabwe with 5% publications each, indicating a less concentrated research effort but with growing interest in GSCM. This distribution highlights the substantial opportunities for increased research and collaboration to enhance the understanding and implementation of green supply chain management practices which can lead to the achievement of sustainability in Africa. Table 9.1 presents the names of journals where the reports included in the final analysis were published.

The statistics on Table 9.1 show that most of the studies published in the scientific journals have a bias towards sustainability. The Management of Environmental Quality, an international journal, was dominant with 15% of the total publications. The second tier which published most of reviews with two each were Journal of Logistic and Operation Management Research, African Journal of Economic and Management Studies, Taylor and Francis, South African Journal of Economic and Management

Table 9.1 Sources of included articles

Journal publication name	*No. of articles*
Journal of Global Operations and Strategic Sourcing	1
Journal of Energy for Sustainable Development	1
Journal of Business & Industrial Marketing	1
Journal of Logistic and Operation Management Research	2
Journal of Scientific African	1
Technological Sustainability Journal	2
Taylor and Francis	2
South African Journal of Economic and Management Sciences	2
Environmental Science and Pollution Research	1
Benchmarking: An International Journal	1
Management of Environmental Quality: An International Journal	3
Journal of Business and Retail Management Research	1
African Journal of Economic and Management Studies	2
Total	20

Source: Secondary Data Analysis (2024)

Sciences, and Technological Sustainability journal. The focus of this research has led to a special analysis of three primary areas among the studies included in the review which are sustainability in general, green supply chain management, and green marketing.

Results and Discussion

The Role of Green Supply Chain Management

The main themes comprised promotion of green innovation, improved supply chain collaboration and relationships, enhanced operational efficiency, enhanced strategic marketing and reputation, and promotion of environmental sustainability. Organisations that seek to implement sustainability can find the themes helpful as they appeal to the roles and effects of GSCM in promoting effective and sustainable supply chain practices in various organisational contexts (Table 9.2).

The Relationship Between GSCM and Green Marketing Agenda

Key findings indicate that there is a relationship between GSCM practices and green marketing outcomes. GSCM promotes the adoption of environmental management systems and sustainability practices such as recycling, reuse, transportation optimisation, and reducing unnecessary inventories which enhance operational efficiency. GSCM also facilitates training and education that foster green innovation and improve sustainability awareness, contributing to sustainable manufacturing. GSCM enhances key green marketing strategies such as green promotion, product, price, and place which lead to improved brand image, customer loyalty, and reduced operational costs. It was also found that organisations that adopt comprehensive GSCM strategies benefit from improved compliance with environmental regulations, better supplier relationships, and positive social impacts which are green marketing outcomes. Studies such as those by Agyemang et al. (2018) and Mafini and Loury-Okoumba (2018) demonstrate that companies in Ghana and South Africa integrating GSCM practices experience enhanced environmental performance, cost savings, and increased market competitiveness through the offering of greener products.

Table 9.2 The main themes on role of green supply chain management

Promotion of green innovation	Improved supply chain collaboration and relationships	Enhanced operational efficiency	Enhanced strategic marketing and reputation	Promotion of environmental sustainability
Improved green absorptive capacity (knowledge acquisition)	Improved green buyer-supplier relationships (governance and trust)	Increased organisational operational efficiency	Improved strategic green marketing orientation	Promotion of environmental sustainability
Improved organisational commitment to the environment	Enhanced environmental participation capabilities	Promotion of sustainable manufacturing	Improved green marketing capabilities	Encourage environmental practices (recycling, reuse, transportation optimisation)
Improved supply chain performance	Enhanced green supplier development capabilities	Enhanced reverse logistics to promote recycling	Promotion of strategic partnerships	Encourage top management commitment to environmental practices
Promotion of green innovation	Improved environmental collaboration with suppliers	Implementation of environmental management systems (EMSs)	Improved corporate reputation	Promotion of green procurement (reducing unnecessary inventories and shortening lead times)
Encourage eco-design of products	Enhanced social sustainability	Efficient warehousing, and reverse logistics	Enhancing green marketing strategies (green promotion, green product, green price, and green place)	Promotion of global attention to climate change
Training and education to foster green innovation	Improved trust within supply chains	Promotion of green distribution/ logistics		Improved cross-border collaborations

Promotion of Green Innovation

Green supply chain management promotes environmentally friendly behaviours like reuse, recycling, and efficient transportation. As it makes training and education targeted at raising sustainability awareness easier, the incorporation of GSCM methods fosters innovation and ongoing learning inside organisations (Nkrumah et al., 2021; Agyabeng-Mensah et al., 2023; Chiwaridzo, 2024). Sustainable manufacturing is crucial for advancing green marketing initiatives, and GSCM plays a major role in enabling this outcome. Businesses that invest in GSCM initiatives like environmentally friendly product design and green processes become more competitive by providing consumers with eco-friendly items that meet their environmental concerns (Famiyeh et al., 2018; Chandran & Bhattacharya, 2021; Boson et al., 2023). In the end, this helps advance the green marketing agenda by increasing consumer demand for environmentally friendly goods.

Improved Supply Chain Collaboration and Relationships

GSCM greatly enhances stakeholder collaboration throughout the supply chain which is also an outcome of green marketing. In this regard, organisations can enhance their relationships with suppliers, consumers, and communities by using sustainable practices. In order to achieve green supply chain practices in developing countries, Ososanmi et al. (2022) established that collaboration between the government, organisations, non-governmental organisations, and individuals is crucial. To ensure that all participants in the supply chain contribute to environmental responsibility, improved coordination is also essential for coordinating supply chain activities with sustainability goals (Mashele, 2018; Jæger et al., 2021; Ososanmi et al., 2022). For instance, Amoako et al. (2023) observed that businesses in Ghana's Cocoa supply chain who adopted GSCM procedures strengthened their connections with suppliers, which in turn improved environmental performance.

Enhanced Operational Efficiency

GSCM has the capacity to increase operational efficiency through waste reduction, inventory optimisation, and lead time reduction (Ali et al., 2021; Masokoameng & Mushavhanamadi, 2023). Research by Ogunlela (2018) has shown that important GSCM techniques like transportation optimisation and minimising needless inventory significantly save operating costs. Research conducted by Mafini and Loury-Okoumba (2018) in South Africa showed that businesses that used green supply chain management strategies were able to reduce costs and increase customer satisfaction by offering more environmentally friendly products. By establishing more efficient and sustainable supply chains, these operational efficiencies do not only have a positive financial impact but also contribute to the greater goal of sustainable marketing (Appiah et al., 2022; Ojo et al., 2022; Ososanmi et al., 2022).

Enhanced Strategic Marketing and Reputation

GSCM is essential for building a business's reputation which is a strategic green marketing agenda. Businesses that use green marketing techniques like green distribution, green pricing, green products, and green promotion are better positioned to develop a strong brand image and win over loyal customers (Ismail, 2023). Agyabeng-Mensah et al. (2018) discovered that businesses in Ghana that implemented GSCM strategies saw increases in their competitiveness and brand recognition. These businesses were able to stand out from the competition and draw in environmentally conscientious customers by coordinating their supply chain operations with sustainability goals which helped them advance their green marketing strategies (Ogunlela, 2018).

Promotion of Environmental Sustainability

The dedication to environmental sustainability as an agenda for green marketing is also at the core of GSCM practices (Charter & Polonsky, 2017; Omuom, 2020). The 2030 Agenda for Sustainable Development of the United Nations and Africa's Agenda 2063 both specify sustainability targets that can be attained through the application of GSCM methods. Mafini and Loury-Okoumba (2018) noted that businesses can help reduce their environmental impact, encourage sustainable resource usage, and promote global environmental sustainability initiatives by incorporating eco-friendly practices into their supply chains. This is crucial in Africa where environmental preservation and sustainable development are directly linked to the promotion of the green marketing agenda. In this sense, GSCM processes are essential in implementing environmental management systems pursued by green marketing (Chung, 2020; Omuom, 2020).

Implications of the Study

The chapter provides an in-depth synthesis of the GSCM body of knowledge in Africa thereby contributing to the theoretical understanding of the role of GSCM to green marketing by highlighting its multifaceted benefits. The chapter offers theoretical framework and guidance for future research by highlighting elements that have not been sufficiently addressed in the literature.

The chapter shows that GCSM adoption and contextual application in Africa is gaining momentum which confirms that sustainability is just not a passing trend but a critical aspect in new product designing, development, and its marketing. The study is therefore a fundamental knowledge platform that policymakers use to create supportive frameworks that encourage GSCM adoption, such as incentives for sustainable practices and stricter environmental regulations.

Businesses should invest in GSCM technology infrastructure, training, and collaboration with marketing to optimise their supply chains for sustainability. The chapter also gives marketing managers a better understanding of various innovations and advantages of incorporating GSCM into their green marketing strategy to promote environmental sustainability. For example, the evaluated studies emphasise the importance of packaging in minimising the environmental impact of both green products and logistics. This means that businesses need to pay close attention to designing packaging that is biodegradable and recyclable while also using fewer material as possible.

Limitations

The study's limitations include a sole focus on the role of GSCM in fostering green marketing agenda. The complexities of GSCM implementation in Africa were not explored by this study to show the challenges which may delay its full adoption and the realisation of sustainability and green marketing outcomes. This suggests that future research should be considered to understand the intricacies of implementing GSCM to promote its widespread adoption across the continent.

Conclusion

The chapter highlights the critical role played by GSCM in advancing sustainable green marketing in Africa by adopting practices that can achieve significant environmental, economic, and social benefits. By demonstrating organisations' dedication to sustainability using GSCM and green marketing practices, they can also improve their standing with stakeholders such as investors, consumers, and environmental policy regulators. This dedication promotes social, economic, and environmental objectives, leading to an organisation's performance that is more sustainable. This chapter also highlights the need for continued research and policy support to overcome barriers and promote widespread adoption of GSCM across the continent. Consequently, the chapter is hoped to provoke future research, and work as reference point to policy makers, academia, and management interested in environmental sustainability in the Global South.

References

Abdel-Baset, M., Chang, V., & Gamal, A. (2019). Evaluation of the green supply chain management practices: A novel neutrosophic approach. *Computers in Industry, 108*(1), 210–220.

Agyabeng-Mensah, Y., Afum, E., Acquah, I. S. K., & Baah, C. (2023). How does supply chain knowledge enhances green innovation? The mediation mechanisms of corporate reputation and non-supply chain learning. *Journal of Business & Industrial Marketing, 38*(4), 852–868.

Agyabeng-Mensah, Y., Afum, E., & Ahenkorah, E. (2020). Exploring financial performance and green logistics management practices: Examining the mediating influences of market, environmental and social performances. *Journal of Cleaner Production, 258*, 120613.

Agyabeng-Mensah, Y., & Tang, L. (2021). The relationship among green human capital, green logistics practices, green competitiveness, social performance and financial performance. *Journal of Manufacturing Technology Management, 32*(7), 1377–1398.

Agyemang, M., Zhu, Q., Adzanyo, M., Antarciuc, E., & Zhao, S. (2018). Evaluating barriers to green supply chain redesign and implementation of related practices in the West Africa cashew industry. *Resources, Conservation and Recycling, 136*, 209–222.

Al-Ghussain, L. (2019). Global warming: Review on driving forces and mitigation. *Environmental Progress & Sustainable Energy, 38*(1), 13–21.

Ali, A., Cao, H., Eid, A., Madkour, T., & Ma, H. (2020). The impact of reverse logistics on environmental sustainability performance. *RIET-IJSET International Journal of Science Engineering and Technology, 8*(2), 18–27.
Ali, E. B., Anufriev, V. P., & Amfo, B. (2021). Green economy implementation in Ghana as a road map for a sustainable development drive: A review. *Scientific African, 12*, e00756.
Amoako, D. K., Zakuan, M. N., Okyere-Kwakye, E., & Tetteh, F. K. (2023). Effect of training and reward on social sustainability in Ghana's Cocoa Supply Chain: The role of green buyer-supplier relationship. *Journal of International Food & Agribusiness Marketing, 35*(2), 212–243.
Appiah, M. K., Odei, S. A., Kumi-Amoah, G., & Yeboah, S. A. (2022). Modeling the impact of green supply chain practices on environmental performance: The mediating role of ecocentricity. *African Journal of Economic and Management Studies, 13*(4), 551–567.
Baeza, R. A. (2018). A methodology for urban sustainability indicator design. *TeMA-Journal of Land Use, Mobility and Environment, 11*(3), 285–303.
Barbier, E. B., & Burgess, J. C. (2017). The sustainable development goals and the systems approach to sustainability. *Economics, 11*(1), 20170028.
Bendig, D., Kleine-Stegemann, L., & Gisa, K. (2023). The green manufacturing framework-A systematic literature review. *Cleaner Engineering and Technology, 13*, 100613.
Berthod, O., Helfen, M., & Sydow, J. (2018). Institutional work for value co-creation: navigating amid power and persistence. *The SAGE handbook of service-dominant logic* (pp. 317–335).
Borges, T. D. M. D., Ganga, G. M. D., Godinho Filho, M., Delai, I., & Santa-Eulalia, L. A. (2021). Development and validation of a cleaner production measurement scale. *Journal of Cleaner Production, 299*, 126907.
Boson, L. T., Elemo, Z., Engida, A., & Kant, S. (2023). Assessment of green supply chain management practices on sustainable venture in Ethiopia. *Logistic and Operation Management Research (LOMR), 2*(1), 96–104.
Caradonna, J. L. (2022). *Sustainability: A history*. Oxford University Press.
Chandran, C., & Bhattacharya, P. (2021). Hotel's best practices as strategic drivers for environmental sustainability and green marketing. In *Consumer behaviour in hospitality and tourism* (pp. 66–81). Routledge.
Charter, M., & Polonsky, M. J. (Eds.). (2017). *Greener marketing: A global perspective on greening marketing practice*. Routledge.
Chen, D., Ignatius, J., Sun, D., Zhan, S., Zhou, C., Marra, M., & Demirbag, M. (2019). Reverse logistics pricing strategy for a green supply chain: A view of customers' environmental awareness. *International Journal of Production Economics, 217*, 197–210.
Chen, Y. S., Huang, A. F., Wang, T. Y., & Chen, Y. R. (2020). Greenwash and green purchase behaviour: The mediation of green brand image and green

brand loyalty. *Total Quality Management & Business Excellence, 31*(1–2), 194–209.

Chiwaridzo, O. T. (2024). Unleashing tomorrow's energy for sustainable development: Pioneering green building technologies and green tourism supply chain management in Zimbabwe's tourism sector. *Energy for Sustainable Development, 78*, 101382.

Chung, K. C. (2020). Green marketing orientation: Achieving sustainable development in green hotel management. *Journal of Hospitality Marketing & Management, 29*(6), 722–738.

Cousins, P. D., Lawson, B., Petersen, K. J., & Fugate, B. (2019). Investigating green supply chain management practices and performance: The moderating roles of supply chain ecocentricity and traceability. *International Journal of Operations & Production Management, 39*(5), 767–786.

Dangelico, R. M., & Vocalelli, D. (2017). "Green Marketing": An analysis of definitions, strategy steps, and tools through a systematic review of the literature. *Journal of Cleaner production, 165*, 1263–1279.

Eriksen, M. B., & Frandsen, T. F. (2018). The impact of patient, intervention, comparison, outcome (PICO) as a search strategy tool on literature search quality: A systematic review. *Journal of the Medical Library Association: JMLA, 106*(4), 420.

Famiyeh, S., Kwarteng, A., Asante-Darko, D., & Dadzie, S. A. (2018). Green supply chain management initiatives and operational competitive performance. *Benchmarking: An International Journal, 25*(2), 607–631.

Faria, J. H. (2014). *Por Uma Teoria Crítica Da Sustentabilidade in a critical theory of sustainability*. Environmental Science Philosophy.

Han, M. S., Hampson, D. P., Wang, Y., & Wang, H. (2022). Consumer confidence and green purchase intention: An application of the stimulus-organism-response model. *Journal of Retailing and Consumer Services, 68*, 103061.

Hashmi, R. (2023). Business performance through government policies, green purchasing, and reverse logistics: Business performance and green supply chain practices. *South Asian Journal of Operations and Logistics, 2*(1), 1–10.

Hinson, R. E., Adeola, O., & Adisa, I. (2021). Introduction: Green marketing and management in emerging markets—Achieving success through people. *Green marketing and management in emerging markets: The crucial role of people management in successful implementation*, 3–11.

Huq, F. A., & Stevenson, M. (2020). Implementing socially sustainable practices in challenging institutional contexts: Building theory from seven developing country supplier cases. *Journal of Business Ethics, 161*(2), 415–442.

Ismail, I. J. (2023). The influence of green dynamic capabilities on brand sustainability among manufacturing enterprises in Tanzania: Mediation of green technology innovation. *Technological Sustainability, 2*(4), 388–403.

Jæger, B., Menebo, M. M., & Upadhyay, A. (2021). Identification of environmental supply chain bottlenecks: A case study of the Ethiopian healthcare supply chain. *Management of Environmental Quality: An International Journal, 32*(6), 1233–1254.

Jalil, E. E. A. (2017). A symbiosis framework in understanding reverse logistics in solid waste. *Advanced Science Letters, 23*(9), 9308–9313.

Jia, F., Zuluaga-Cardona, L., Bailey, A., & Rueda, X. (2018). Sustainable supply chain management in developing countries: An analysis of the literature. *Journal of Cleaner Production, 189*, 263–278.

Kauppi, K., & Hannibal, C. (2017). Institutional pressures and sustainability assessment in supply chains. *Supply Chain Management: An International Journal, 22*(5), 458–472.

Khan, S. A. R., Qianli, D., SongBo, W., Zaman, K., & Zhang, Y. (2017). Environmental logistics performance indicators affecting per capita income and sectoral growth: Evidence from a panel of selected global ranked logistics countries. *Environmental Science and Pollution Research, 24*, 1518–1531.

Lammers, J. C., Garcia, M. A., Putnam, L. L., & Mumby, D. K. (2014). Institutional theory. *The SAGE handbook of organizational communication: Advances in theory, research, and methods*, 195–216.

Maditati, D. R., Munim, Z. H., Schramm, H. J., & Kummer, S. (2018). A review of green supply chain management: From bibliometric analysis to a conceptual framework and future research directions. *Resources, Conservation and Recycling, 139*, 150–162.

Mafini, C., & Loury-Okoumba, W. V. (2018). Extending green supply chain management activities to manufacturing small and medium enterprises in a developing economy. *South African Journal of Economic and Management Sciences, 21*(1), 1–12.

Mancin, S., Sguanci, M., Anastasi, G., Godino, L., Cascio, A. L., Morenghi, E., ... & De Marinis, M. G. (2024). A methodological framework for rigorous systematic reviews: Tailoring comprehensive analyses to clinicians and healthcare professionals. Methods.

Mashele, F. (2018). *Supply chain management predictors of sustainable procurement and inclusive business in South Africa* (Doctoral dissertation).

Masokoameng, B., & Mushavhanamadi, K. (2023). Implementation of green supply chain management practices in South African Manufacturing Industry.

Methley, A. M., Campbell, S., Chew-Graham, C., McNally, R., & Cheraghi-Sohi, S. (2014). PICO, PICOS and SPIDER: A comparison study of specificity and sensitivity in three search tools for qualitative systematic reviews. *BMC Health Services Research, 14*(1), 1–10.

Moher, D., Shamseer, L., Clarke, M., Ghersi, D., Liberati, A., Petticrew, M., ... & Prisma-P Group. (2015). Preferred reporting items for systematic review and meta-analysis protocols (PRISMA-P) 2015 statement. *Systematic Reviews, 4*, 1–9.

Mwaura, A. W., Letting, N., Ithinji, G. K., & Bula, H. O. (2016). Green distribution practices and competitiveness of food manufacturing firms in Kenya.

Namagembe, S., Ryan, S., & Sridharan, R. (2019a). Green supply chain practice adoption and firm performance: Manufacturing SMEs in Uganda. *Management of Environmental Quality: An International Journal, 30*(1), 5–35.

Nguyen, A. T., Parker, L., Brennan, L., & Lockrey, S. (2020). A consumer definition of eco-friendly packaging. *Journal of Cleaner Production, 252*, 119792.

Nkrumah, S. K., Asamoah, D., Annan, J., & Agyei-Owusu, B. (2021). Examining green capabilities as drivers of green supply chain management adoption. *Management Research Review, 44*(1), 94–111.

Ogunlela, G. O. (2018). Green supply chain management as a competitive tool in the fast-moving consumer goods manufacturing industry. *Journal of Business and Retail Management Research, 12*(4). https://doi.org/10.24052/JBRMR/V12IS04/ART-17

Ojo, L. D., Adeniyi, O., Ogundimu, O. E., & Alaba, O. O. (2022). Rethinking green supply chain management practices impact on company performance: A close-up insight. *Sustainability, 14*(20), 13197.

Omuom, S. A. (2020). *Promoting enterprise sustainability through Green Marketing Models: A case study of tissue manufacturing at Chandaria Industries Limited Nairobi, Kenya* (Doctoral dissertation, University of Nairobi).

Ouzzani, M., Hammady, H., Fedorowicz, Z., & Elmagarmid, A. (2016). Rayyan—a web and mobile app for systematic reviews. *Systematic Reviews, 5*, 1–10.

Ososanmi, A. O., Ojo, L. D., Ogundimu, O. E., & Oke, A. E. (2022). Drivers of green supply chain management: A close-up study. *Environmental Science and Pollution Research, 29*(10), 14705–14718.

Page, M. J., McKenzie, J. E., Bossuyt, P. M., Boutron, I., Hoffmann, T. C., Mulrow, C. D., ... & Moher, D. (2021). The PRISMA 2020 statement: an updated guideline for reporting systematic reviews. bmj, 372.

Pahlevi, M. R., & Suhartanto, D. (2020). The integrated model of green loyalty: Evidence from eco-friendly plastic products. *Journal of Cleaner Production, 257*, 120844.

Pang, R., & Zhang, X. (2019). Achieving environmental sustainability in manufacture: A 28-year bibliometric cartography of green manufacturing research. *Journal of Cleaner Production, 233*, 84–99.

Patagonia. (2021). Retrieved August 2024, from https://www.patagonia.com/on/demandware.static/-/Library-Sites-PatagoniaShared/default/dw18ad9c7c/PDF-US/Patagonia-2021-BCorp-Report-Updated-2-15-22.pdf

Patagonia. (2024). https://www.patagonia.com/environmental-responsibility-materials/. (Accessed 08/2024).

Patino, C. M., & Ferreira, J. C. (2018). Inclusion and exclusion criteria in research studies: definitions and why they matter. *Jornal Brasileiro de Pneumologia, 44*, 84–84.

Peters, B. G. (2022). Institutional theory. In *Handbook on theories of governance* (pp. 323–335). Edward Elgar Publishing.

Pollock, A., & Berge, E. (2018). How to do a systematic review. *International Journal of Stroke, 13*(2), 138–156.

Poveda, C. A. (2017). The theory of dimensional balance of needs. *International Journal of Sustainable Development & World Ecology, 24*(2), 97–119.

Sarkodie, S. A. (2018). The invisible hand and EKC hypothesis: What are the drivers of environmental degradation and pollution in Africa? *Environmental Science and Pollution Research, 25*(22), 21993–22022.

Seman, N. A. A., Govindan, K., Mardani, A., Zakuan, N., Saman, M. Z. M., Hooker, R. E., & Ozkul, S. (2019). The mediating effect of green innovation on the relationship between green supply chain management and environmental performance. *Journal of Cleaner Production, 229*, 115–127.

Sharma, N. K., Kumar, V., Verma, P., & Luthra, S. (2021). Sustainable reverse logistics practices and performance evaluation with fuzzy TOPSIS: A study on Indian retailers. *Cleaner Logistics and Supply Chain, 1*, 100007.

Starik, M., & Kanashiro, P. (2013). Toward a theory of sustainability management: Uncovering and integrating the nearly obvious. *Organization & Environment, 26*(1), 7–30.

Statista. (2020). Environmentally friendly brand attributes that are important to consumers worldwide in 2020.

Wang, C., Zhang, Q., & Zhang, W. (2020). Corporate social responsibility, Green supply chain management and firm performance: The moderating role of big-data analytics capability. *Research in Transportation Business & Management, 37*, 100557.

Waring, T. M. (2016). An evolutionary approach to sustainability science. *Cliodynamics, 7*(1), 119–167.

Yang, C. L., Lin, S. P., Chan, Y. H., & Sheu, C. (2010). Mediated effect of environmental management on manufacturing competitiveness: An empirical study. *International Journal of Production Economics, 123*(1), 210–220.

Yildiz Çankaya, S., & Sezen, B. (2019). Effects of green supply chain management practices on sustainability performance. *Journal of Manufacturing Technology Management, 30*(1), 98–121.

CHAPTER 10

Green Reverse Logistics as a Sustainable Strategy towards Circular Economies in Africa

Shakerod Munuhwa and *Dumisani Mawonde*

Introduction

Globally, many countries currently focus on creating a circular economy to overcome the challenge of unsustainable manufacturing and consumption processes that cause damage to the environment and natural ecosystems (Jayarathna et al., 2024). The inception of the United Nations (UN) Sustainable Development Goals (SDGs) directed many countries to focus on efficient production and sustainable economic activities that do not contribute to environmental damage in order to achieve a circular

S. Munuhwa (✉)
Department of Management, Glasgow School for Business and Society, Glasgow Caledonian University, Glasgow, UK

D. Mawonde
Department of Logistics, Faculty of Business and Economic Science, Nelson Mandela University, Gqeberha, South Africa
e-mail: s227139 51@mandela.co.za

A. Muposhi, B. Nyagadza (eds.), *Sustainable Green Marketing Strategies for a Circular Economy*, Palgrave Studies of Marketing in Emerging Economies,
https://doi.org/10.1007/978-3-031-81627-7_10

economy status (Ritter et al., 2024). According to the World Economic Forum (WCF) (2021), creating and developing a circular economy offers countries the opportunity to improve the standards of living of people and to achieve sustainable economic development.

Many countries have put in place strategies towards a circular economy in line with the SDGs (Koech et al., 2023). Literature indicates that circular economies include the practice of harnessing and efficient utilisation of resources and the practice of sensible waste management to fulfil the requirements of people and achieve several SDGs. Singh and Ordonez (2016) cited in Mallick et al. (2023) posit that reverse logistics is increasingly receiving attention as a strategy towards a circular economy in many countries where the focus is on recycling and reuse as means of recovering goods, parts, and materials. The importance of reverse logistics is that it provides countries with the opportunity to benefit from the efficient utilisation of resources through reusing and recycling End of Use (EoU) and End of Life (EoL) products. Therefore, governments now emphasise companies to adopt and implement reverse logistics practices in order to attain circular economic status (Mallick et al., 2023). Governments have also put in place laws and regulations to ensure that companies and other key players in economic activities adopt reverse logistics as a driver to move towards a circular economy.

Many countries still use the traditional linear economy model which is characterised by a "take, make, dispose" approach. The problem with the traditional linear economy model is that it has caused major resource depletion and environmental devastation in many countries. The World Bank (2018) pointed out that the global waste production is anticipated to reach 3.4 billion tonnes in 2050; therefore, effective reverse logistics practices are essential for reducing this environmental burden. By integrating green reverse logistics (GRL) practices, companies can minimise waste, reduce their carbon footprint, and conserve natural resources. For instance, the European Union reported that implementing circular economy strategies could reduce the continent's carbon emissions by 450 million tonnes annually by 2030 (Ellen MacArthur Foundation, 2015). Examples of developed countries that have successfully achieved circular economy status through green reverse logistics include China, Japan, Germany, and the Netherlands (Jayarathna et al., 2024). These countries have encouraged putting reverse logistics into practice in companies across sectors such as manufacturing, fashion, construction, and textile.

In developing countries, particularly in Africa, the implementation of sustainable reverse logistics practices presents unique challenges. Despite the continent's rich natural resources, it faces significant infrastructural, economic, and regulatory hurdles. The World Bank notes that only 9% of waste in Sub-Saharan Africa is recycled, compared to 25–30% in developed regions (World Bank, 2018). This disparity is attributed to inadequate waste management infrastructure, limited technical expertise, and insufficient financial investment. Additionally, informal waste management systems dominate many African countries, which often lack the regulatory frameworks to support formal GRL practices. For example, Nigeria's informal sector handles over 60% of its waste, leading to inefficient recycling processes and environmental pollution (Ogunseye et al., 2020). Addressing these challenges requires comprehensive policy reforms, investment in infrastructure, and capacity building to enable the successful implementation of GRL in Africa.

This research focuses on green reverse logistics as a strategy towards the creation and development of circular economies in developing countries in Africa. A novel approach to reverse logistics called "green reverse logistics" seeks to maximise resource efficiency and waste management. The significance of this research is that it focuses on one crucial technique for the establishment and growth of a circular economy in African countries, with a particular focus on recyclable products. The adoption of green reverse logistics practices in Africa is critical for achieving broader sustainable development goals. Effective GRL can drive economic growth, create job opportunities, and promote environmental sustainability. For instance, the circular economy has the potential to generate $4.5 trillion in economic benefits globally by 2030, with significant opportunities for emerging markets like Africa (Khan & Mihaisi, 2023). Moreover, fostering sustainable practices in reverse logistics can enhance Africa's global competitiveness by aligning with international environmental standards and attracting foreign investment. Implementing GRL strategies can also help mitigate climate change impacts as waste management contributes to about 5% of global greenhouse gas emissions (Gautam & Agrawal, 2021). Therefore, prioritising sustainable reverse logistics in Africa is not only vital for environmental preservation but also for socio-economic development and resilience against global challenges. The significance of this research is that it is expected to come up with recommendations that will help African countries to focus on green reverse logistics as an important strategy towards a circular economy in order to achieve sustainable economic development in line with SDGs. Additionally, the results of this study add to the corpus of knowledge in the field of green reverse logistics and its part in establishing a circular economy contribution and the achievement of sustainable economic development. Therefore, this research is expected to address the following primary objective: to examine the role of green reverse logistics practices in the transition towards circular economies in Africa.

The next chapter sections are organised as follows: The "Theories Underpinning the Research" section reviews the literature that is related to this research; third is the methodology; fourth looks at the findings of this research whilst fifth looks at the conclusion, recommendations, managerial implications, implications for future research, and limitations of the research.

Theories Underpinning the Research

The section covers theories underpinning the study. This research is guided by the Triple Bottom Line (TBL) framework and the Resource Based View (RBV) theories as discussed below.

The Resource-Based View (RBV)

The RBV is a strategic management theory that emphasises a firm's internal resources as the primary source of sustainable competitive advantage. According to Barney (1991), the resources must be Valuable, Rare, Inimitable, and Non-substitutable (VRIN) to confer a competitive edge. These resources encompass physical assets, human capital, organisational capabilities, and intellectual property. The RBV shifts the focus from external market conditions to internal capabilities, highlighting the critical role of resource management in achieving strategic objectives.

The application of the RBV to green reverse logistics (GRL) highlights the significance of internal resources in developing sustainable logistics practices. GRL encompasses activities such as recycling, remanufacturing, and proper disposal of products, all of which require specialised capabilities and infrastructure. Firms with advanced technological capabilities, skilled human resources, and efficient logistical networks are better positioned to implement GRL practices effectively. For example, companies with robust IT systems can better track and manage reverse logistics operations, ensuring efficient recovery and recycling of materials (Mugoni et al., 2023). Leveraging these unique resources enables firms to reduce environmental impact, comply with regulatory requirements, and enhance their reputation for sustainability, thus gaining a competitive advantage.

As well, the application of RBV to GRL is particularly pertinent for promoting circular economies in Africa. The continent faces significant challenges in waste management and resource recovery due to infrastructural and regulatory constraints. However, firms that can harness their internal resources effectively can turn these challenges into opportunities. For instance, companies with innovative technologies and skilled personnel can develop effective GRL systems, even in resource-constrained environments. A notable example is the South African recycling industry, which has developed sophisticated systems for recovering and repurposing waste materials despite limited infrastructure (Godfrey & Oelofse, 2017). By investing in capabilities such as advanced recycling technologies and training programmes for employees, firms can enhance their GRL operations and contribute to the circular economy (Gobena & Azene, 2022).

In Africa, the RBV framework can guide firms in identifying and developing the critical resources needed for GRL. Investing in local talent development can create a skilled workforce capable of managing complex reverse logistics processes. Partnerships with local communities and governments can also enhance access to resources and infrastructure, facilitating the implementation of GRL practices. For example, the partnership between Coca-Cola Beverages Africa and local governments to improve plastic bottle recycling demonstrates how leveraging internal and external resources can drive sustainable practices (Coca-Cola Beverages Africa, 2020). By focusing on internal resource development, African firms can overcome external challenges and contribute to a sustainable circular economy.

Despite its strengths, the RBV is not without criticism. Critics argue that the RBV tends to overlook the dynamic nature of external environments, which can significantly impact a firm's resources and capabilities (Helfat, et al., 2023). This constraint is especially pertinent in light of Africa, where external factors such as political instability, economic volatility, and regulatory changes can influence the effectiveness of GRL practices. Additionally, the RBV may not adequately address the complexities of resource integration and coordination in diverse and fragmented supply chains, common in many African countries.

Moreover, the RBV's emphasis on internal resources might lead firms to under-appreciate the importance of external collaborations and partnerships in achieving sustainability goals. For example, while internal capabilities are crucial, the success of GRL initiatives often depends on cooperation with external stakeholders such as suppliers, customers, and regulatory bodies. The informal waste sector in many African countries exemplifies this, where informal recyclers play a critical role in waste recovery processes. Integrating these external actors into formal GRL systems requires more than just internal resources; it demands strategic partnerships and policy support (Ozdemir et al., 2023).

While the RBV provides a valuable framework for comprehending how internal resources function in GRL, it must be complemented with an appreciation of external factors and collaborations to fully address the difficulties and possibilities of applying environmentally friendly logistics techniques.

The Triple Bottom Line (TBL)

The TBL framework was introduced by John Elkington (1994). This theory extends the customary emphasis on profit (economic value) to include social and environmental dimensions. This holistic approach assesses a company's performance using the three main criteria of profit, planet, and people. The TBL framework encourages businesses to consider the full impact of their operations and strive for sustainability by balancing financial performance with social equity and environmental responsibility.

Applying the TBL framework to green reverse logistics (GRL) practices in Africa presents a comprehensive strategy for evaluating reverse logistics operations' sustainability. GRL, which includes processes such as recycling, remanufacturing, and waste disposal, inherently aligns with the environmental pillar of the TBL by reducing waste and promoting resource efficiency. For example, effective GRL practices can significantly lower the environmental footprint of manufacturing industries in Africa by ensuring that materials are reused and recycled, thereby reducing the need for virgin resources (Govindan et al., 2022).

From a social perspective, GRL can create job opportunities and improve working conditions in the recycling and waste management sectors. For instance, integrating informal waste pickers into formal recycling programmes not only improves their livelihoods but also enhances the efficiency and scope of GRL initiatives (Godfrey & Oelofse, 2017). Furthermore, companies that adopt GRL practices can enhance their corporate social responsibility (CSR) profiles, fostering community goodwill and stakeholder engagement.

Economically, GRL can result in cost reductions and new sources of income. By recovering valuable materials from returned products, companies can reduce raw material costs and potentially generate income from selling refurbished goods. For example, the adoption of GRL practices in South Africa's automotive industry has led to cost reductions and improved resource efficiency, contributing to both economic and environmental sustainability (Barnes & Morris, 2008).

The TBL framework complements the Resource-Based View (RBV) by providing a broader perspective on sustainability that encompasses external impacts and stakeholder engagement. As the RBV concentrates on leveraging resources within the organisation to achieve viable advantage, the TBL ensures that these resources are managed in a way that balances economic performance with social and environmental responsibilities. This dual approach is particularly valuable for GRL in Africa, where external factors such as community involvement and regulatory compliance are crucial for the success of sustainable initiatives (Govindan et al., 2022).

Integrating TBL with RBV allows firms to develop a more robust strategy for GRL. For example, a company with strong internal capabilities in logistics and recycling (RBV) can use the TBL framework to ensure that these capabilities are applied in a way that benefits the environment and society. This integrated approach can enhance the firm's reputation, meet regulatory requirements, and attract socially conscious consumers, thereby creating a comprehensive sustainable strategy (Godfrey & Oelofse, 2017).

Despite its strengths, the TBL framework has several criticisms. One major critique is its lack of a clear, standardised measurement system, which makes it difficult to quantify and compare the social and environmental impacts of different companies (Srivastava et al., 2022). This lack of standardisation can lead to inconsistencies and challenges in implementing TBL effectively, particularly in the diverse and complex contexts found in Africa.

Additionally, critics argue that the TBL framework can sometimes lead to superficial compliance rather than genuine sustainability efforts. Companies might engage in "greenwashing" by promoting minor environmental initiatives while neglecting more substantial changes needed to achieve true sustainability (Ibid.). This issue is particularly pertinent in Africa, where regulatory enforcement may be weaker and companies might exploit the TBL framework to enhance their public image without making significant environmental or social improvements.

Literature Review

This literature review synthesises and examines research on the identification of green reverse logistics practices and their contributions to circular economies. The following subsection explores the concept of green reverse logistics in detail.

The Concept of Green Reverse Logistics

Reverse logistics refers to the processes and activities involved in managing the return, exchange, repair, recycling, and disposal of products and materials within a supply chain or distribution network. Unlike forward logistics, which deals with the movement of products from manufacturers to consumers, reverse logistics deals with the flow of goods in the opposite direction, typically from consumers back to manufacturers or suppliers (Bag & Gupta, 2019). The primary objectives of reverse logistics include recovering value from returned products, reducing costs associated with returns and disposal, and managing end-of-life products in an efficient and responsible manner (Derse, 2024; Butt et al., 2023).

Green reverse logistics (GRL) refers to the application of environmentally sustainable practices and principles within the processes of reverse logistics. Green reverse logistics focuses on integrating environmental considerations into the processes of reverse logistics, which traditionally encompass the management of product returns, exchanges, repairs, and recycling within supply chains. Unlike conventional reverse logistics, which primarily aims to recover value or reduce costs associated with returned products, green reverse logistics places a strong emphasis on sustainability principles. This includes reducing carbon footprints, minimising waste generation, and promoting the reuse or recycling of materials to enhance environmental performance (Bag & Gupta, 2019).

In contrast to traditional reverse logistics, where the primary goals are cost reduction and operational efficiency, green reverse logistics prioritises environmental conservation and sustainable resource management. This approach requires businesses to adopt practices such as eco-friendly packaging, efficient transportation routes to reduce emissions, and the implementation of recycling programs that comply with environmental regulations. By integrating these green practices into reverse logistics operations, companies not only mitigate environmental impacts but also enhance their corporate social responsibility (CSR) profiles and appeal to environmentally conscious consumers (Derse, 2024).

Furthermore, green reverse logistics contributes to broader sustainability goals by fostering circular economy principles. This involves designing products with end-of-life considerations in mind, enabling easier disassembly and material recovery. By aligning economic incentives with environmental stewardship, businesses can achieve long-term benefits such as reduced resource consumption, lower waste disposal costs, and improved brand reputation as leaders in sustainable business practices. Embracing green reverse logistics thus represents a strategic approach for companies looking to enhance both environmental sustainability and operational efficiency across their supply chains (Fernando et al., 2022).

Green reverse logistics (GRL) integrates several essential components aimed at promoting environmental sustainability throughout the reverse logistics process. Green procurement is pivotal, involving sourcing products and materials that adhere to stringent environmental standards and support sustainable practices. For instance, in Africa, initiatives like the Sustainable Procurement Program by the United Nations Environment Programme (UNEP) encourage organisations to procure goods and services that minimise environmental impact and promote social responsibility. This approach not only supports local economies but also ensures products meet international sustainability criteria, contributing to broader environmental goals (UNEP, 2020).

Eco-design is another critical component of GRL, focusing on designing products with minimal environmental impact. In Africa, initiatives promoting eco-design include projects aimed at developing sustainable agricultural machinery and equipment that reduce energy consumption and promote efficient use of resources. For example, organisations like the African Circular Economy Alliance (ACEA) collaborate with manufacturers to design agricultural tools using recycled materials and energy-efficient technologies, thus supporting sustainable practices across the continent (ACEA, 2023).

Product recovery within GRL entails strategies for reclaiming value from returned or end-of-life products. In Africa, innovative approaches include the establishment of reverse logistics networks for electronic waste (e-waste), where organisations collect, refurbish, and recycle electronic devices to recover valuable materials like metals and reduce environmental pollution. For example, the Waste Electrical and Electronic Equipment (WEEE) Centre in Kenya partners with local businesses to responsibly manage e-waste, contributing to sustainable development by creating jobs and promoting environmental stewardship (WEEE Centre, 2021).

Waste management forms a crucial part of GRL in Africa, involving the implementation of sustainable waste disposal practices. Projects such as community-based recycling initiatives in South Africa empower local communities to collect and segregate waste for recycling, reducing landfill waste and promoting resource recovery. Organisations like WastePlan collaborate with municipalities to establish efficient waste management systems that comply with environmental regulations and support sustainable development goals (Scrioşteanu & Criveanu, 2024). By integrating these components into GRL strategies, African countries can enhance environmental sustainability, promote economic development, and foster resilient supply chains that align with global sustainability objectives.

Several drivers propel the adoption of green reverse logistics (GRL) within contemporary supply chains, reflecting a growing awareness and commitment to environmental sustainability. One significant driver is regulatory pressures and compliance requirements imposed by national and international environmental legislation. For instance, directives from the European Union (EU) such as the Waste Framework Directive and the Circular Economy Action Plan set stringent targets for waste reduction, recycling rates, and resource efficiency (European Commission, 2020). These regulations compel businesses to integrate green practices into their reverse logistics operations to avoid penalties and align with legal obligations, thereby driving the adoption of GRL.

Consumer preferences and market demands also serve as influential drivers for GRL. Increasingly, consumers are prioritising products and services from companies that demonstrate environmental responsibility and sustainability. This shift in consumer behaviour compels businesses to adopt GRL practices to meet market expectations, enhance brand reputation, and maintain competitive advantage (Chartered Institute of Procurement and Supply, 2021). Companies that embrace green reverse logistics not only cater to eco-conscious consumers but also differentiate themselves in the marketplace by demonstrating their commitment to sustainability.

Furthermore, cost savings and operational efficiencies derived from GRL initiatives provide another compelling driver for adoption. Optimising processes such as product recovery, recycling, and waste management enables businesses to reduce operational costs associated with waste disposal and raw material acquisition. For instance, implementing efficient reverse logistics systems can lead to savings in transportation, storage, and disposal costs, while also generating revenue from recovered materials

(Mayanti & Helo, 2024). These economic incentives encourage organisations to invest in GRL practices as a means to achieve both environmental and financial sustainability. The following section reviews the importance of the circular economy.

The Importance of a Circular Economy

Aithal and Aithal (2023) opine that a circular economy is a type of economy where economic growth is decoupled from resource consumption as well as damage to the environment. With a circular economy, the industry optimises the use of resources by minimising waste, maximising the utilisation of resources, and promoting closed-loop systems. This research focuses on green reverse logistics as a strategy for creating a closed-loop economic system where scarce resources are not wasted.

Since many countries have committed to the achievement of SDGs the world over, a circular economy is important because it helps in creating prosperity for people and the preservation of the environment (Mingst et al., 2022). Therefore, a circular economy enables a country to focus on resource efficiency and waste minimisation (Ritter et al., 2024). Also, due to the scarcity of resources and as many natural resources in African countries are depleted due to massive exploitation by humans and the effects of climate change, circular economies are key to managing and preserving the available resources for continued use by future generations.

The concept of a circular economy encompasses a range of benefits that significantly contribute to sustainable development, resource efficiency, and environmental stewardship. One of the primary advantages is resource conservation. By promoting practices such as recycling, remanufacturing, and refurbishing, a circular economy reduces the reliance on virgin materials. This conservation of natural resources, including minerals, water, and energy, helps mitigate environmental degradation associated with resource extraction and eases the pressures on ecosystems (Geissdoerfer et al., 2017). The emphasis on reusing materials and products conserves finite resources, ensuring their availability for future generations and promoting a more sustainable approach to consumption and production.

Another positive central to the circular economy is the idea of waste minimisation through the design of products that can be easily repaired, reused, or recycled. This practice significantly reduces the volume of waste sent to landfills or incinerators, addressing pollution and improving overall environmental quality (Govindan et al., 2022). By focusing on the entire lifecycle of products, the circular economy aims to eliminate waste at every stage, fostering a more efficient use of materials. Additionally, transitioning to a circular economy creates new economic opportunities, fostering job creation in sectors such as recycling, repair services, and sustainable manufacturing. It stimulates entrepreneurship and investment in technologies that support circular practices, thereby enhancing competitiveness and resilience in the global market (Jayarathna et al., 2024).

Moreover, a circular economy contributes to climate change mitigation by closing material loops and reducing energy-intensive processes associated with extraction and production. This approach lowers greenhouse gas emissions and enhances overall environmental sustainability by minimising the carbon footprint of industrial and consumer activities (Koech et al., 2023). Products designed for longevity and ease of repair contribute to consumer satisfaction and reduce the need for frequent replacements, promoting sustainable consumption patterns and fostering a culture of responsible product stewardship.

Furthermore, circular economy practices enhance resilience against raw material scarcity and price volatility by diversifying supply chains. By utilising secondary resources and maintaining material flows within local or regional economies, businesses can reduce dependency on global supply chains, mitigating risks associated with geopolitical instability or disruptions (Makaleng & Hove-Sibanda, 2022). Embracing circular economy principles also aligns with increasingly stringent environmental regulations and sustainability standards, enhancing corporate social responsibility profiles and strengthening stakeholder trust through a demonstrated commitment to sustainable practices and ethical business conduct.

Contribution of GRL Practices to Circular Economy

GRL is widely recognised for its potential to advance the circular economy through the recovery, reuse, and recycling of materials, which contributes to resource efficiency and waste minimisation. However, while there is broad agreement on the fundamental benefits of GRL, there are also highlights of divergent views regarding its implementation and impact across different contexts and industries.

One area of contribution noted is the role of GRL practices in extending product life cycles, which is seen as essential for minimising the effects of production and consumption on the environment. Authors such as Mallick et al. (2023) and Geissdoerfer et al. (2017) emphasise that practices like remanufacturing and refurbishing, enabled by GRL, are critical for keeping products in use longer and thereby decoupling economic growth from resource depletion. This aligns with the circular economy's goal of reducing reliance on virgin materials. However, there is a gap identified regarding the specific challenges associated with implementing these practices across different sectors, particularly in industries where product design and manufacturing processes are not conducive to easy remanufacturing or refurbishment (Gharibi & Abdollahzadeh, 2021). This suggests a need for more sector-specific research to explore how GRL can be tailored to different industrial contexts.

Studies also converge on the importance of GRL for enhancing resource efficiency through material recovery and recycling. Khan et al. (2017) argue that GRL practices are crucial for maintaining the flow of materials within the economy, thereby reducing the need for new resource extraction. This perspective is supported by Gharibi & Abdollahzadeh (2021), who highlight the role of GRL in reducing environmental pressures associated with raw material extraction. However, while these benefits are widely acknowledged, there is divergence in the effectiveness of GRL practices in different geographic and regulatory contexts. For instance, the effectiveness of GRL practices in developing countries, where infrastructure for recycling and material recovery may be less developed, is underexplored (Ritter et al., 2024). This represents a significant gap, indicating a need for research that addresses how GRL can be adapted to the specific challenges and opportunities in different regions, particularly in the Global South.

A recurring theme in the studies is the role of GRL in waste reduction, which is central to the circular economy's objective of minimising waste generation. Authors such as Gobena and Azene (2022) and Khan et al. (2017) discuss how GRL practices, by diverting products from landfills and reintroducing them into the production cycle, contribute to significant waste reduction. However, the literature often lacks a nuanced discussion of the potential trade-offs associated with GRL, such as the energy and resources required for the logistics and processing involved in material recovery. This suggests a need for more comprehensive assessments that consider the full environmental impacts of GRL practices rather than focusing solely on their waste reduction benefits (Ofori, 2023).

Economic benefits associated with GRL, such as cost savings and job creation, are another area of convergence in the literature. The Ellen MacArthur Foundation (2015) discusses the potential for GRL to create new business opportunities and enhance the competitiveness of firms by reducing production costs through material recovery. This economic perspective is widely supported, yet there is a divergence in views regarding the scalability of these benefits across different industries and regions. While some authors highlight the potential for widespread adoption of GRL practices, others caution that the economic viability of GRL may be limited in industries where the cost of recovery and remanufacturing is high (Epoh & Mafini, 2018). This divergence suggests a need for more research into the economic conditions and policy frameworks that can support the broader implementation of GRL practices.

Methodology

The researchers have adopted the Systematic Literature Review and PRISMA techniques for this particular study (Page et al., 2021). The PRISMA technique consists of four steps which are depicted in a flowchart in Fig. 10.1, which shows the process of gathering original documents to be visualised: identification, screening, assess, and synthesise. The researchers were guided by Correa-Vaca et al. (2024) on the eligibility criteria for the review of green reverse logistics (GRL) practices, addressing circular economy principles, published in peer-reviewed journals or conference proceedings from 2010 to 2024, and available in English. Information sources comprised databases such as PubMed, Scopus, Web of Science, and Google Scholar. A mix of keywords was used in the search approach, including "green reverse logistics," "circular economy," "sustainability," and "Africa," with Boolean operators (AND, OR) to refine searches. Study selection involved two reviewers independently screening titles, abstracts, and author key words, followed by full-text assessments, with disagreements settled by conversation or a third reviewer. Data extraction utilised a standardised form to capture study characteristics, GRL practices, CE contribution, or benefits and these were verified by a second reviewer. Data synthesis included meta-analysis to quantitatively identify common patterns, themes, and gaps, grouping studies based on key GRL practices and their contributions to circular economies.

Identification

A thorough search was carried out on several databases, including Google Scholar, PubMed, Scopus, and Web of Science, resulting in 453 records.

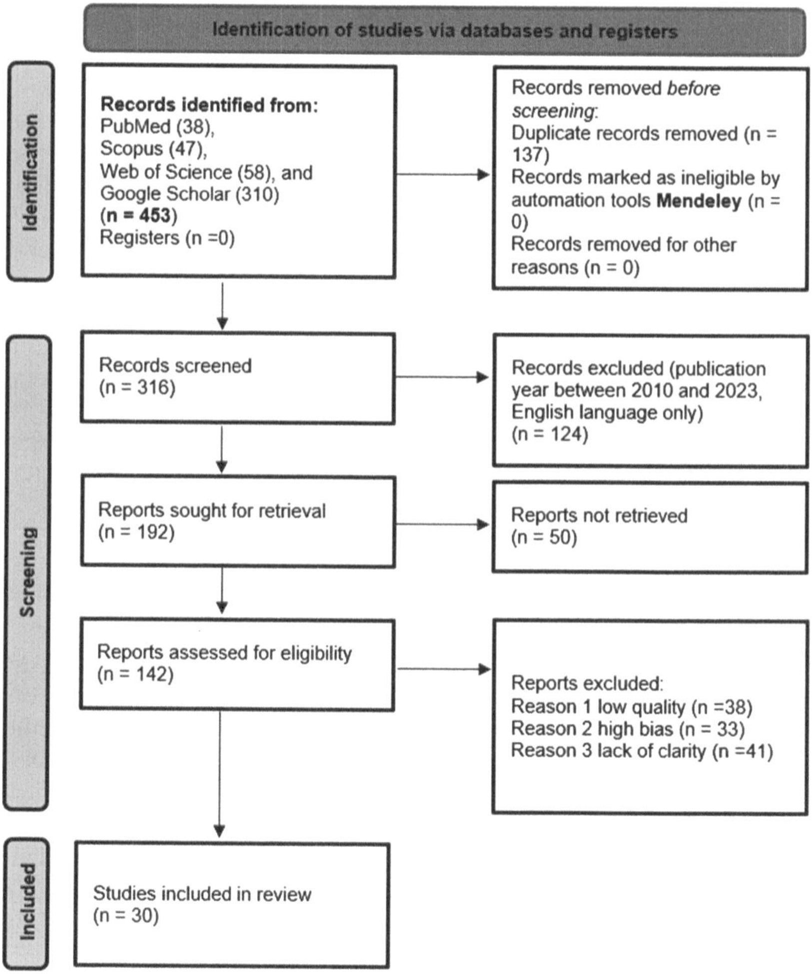

Fig. 10.1 PRISMA Flow. (Source: Adapted from Page et al. (2021) and edited)

This approach was chosen to gather a broad range of studies, enhancing the reliability and depth of the review. Furthermore, 30 records were identified through citation searching, manual searches of key journals, and consultations with experts.

Screening

To avoid redundancy, the manual process was employed to identify and remove duplicate records, resulting in the elimination of 150 duplicates. This step ensures that each study is considered only once, maintaining the integrity of the review process. This initial screening helps streamline the review process, allowing researchers to focus on more pertinent studies. Additionally, 10 records were removed due to language barriers (non-English), lack of access to full text, or irrelevance after an initial title and abstract review, ensuring that only accessible and relevant studies were considered.

After these exclusions, 320 records remained and were subjected to a detailed screening of titles and abstracts, ensuring that only studies relevant to the research question were retained. Of these, 220 records were excluded due to irrelevance to the research question, a focus outside Africa, or a lack of alignment with GRL and circular economy principles. This stringent exclusion criterion ensures that the final set of studies is directly pertinent to the topic.

Studies Included

The final set of 32 studies that met all inclusion criteria provides robust and relevant data on the impact of green reverse logistics (GRL) practices towards circular economies in Africa. These studies form the basis for the meta-analysis, offering in-depth insights into the effectiveness and implementation of GRL practices. By focusing on these 32 studies, the review ensures a concentrated and detailed analysis of the most pertinent research available, highlighting successful strategies and identifying common challenges within the African context.

Findings

Figure 10.2 shows the findings on GRL practices and their contribution towards a circular economy.

Figure 10.2 identifies recycling and material recovery, stakeholder collaboration, and reverse logistics networks as the most frequently studied and implemented GRL practices. These practices are critical for advancing circular economy principles, demonstrating their pivotal role in sustainable supply chain management. On the CE contributions side, resource

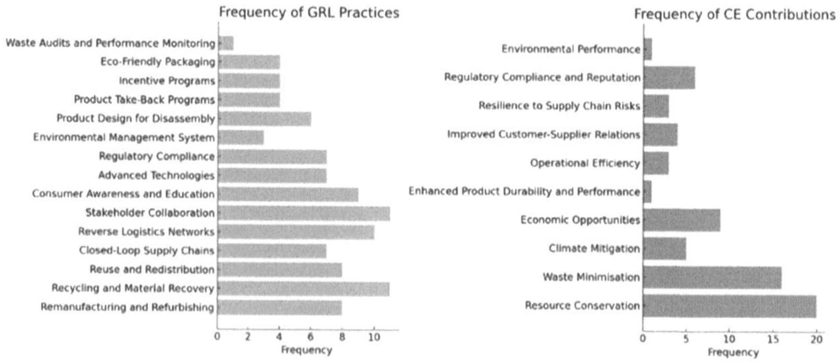

Fig. 10.2 GRL practice and CE contributions observed. (Source: Secondary data (2024))

conservation and waste minimisation emerge as the most common benefits associated with GRL practices. These contributions are essential for achieving sustainability goals, the significance of waste minimisation and resource efficiency in circular economy initiatives. Other significant contributions include economic opportunities and regulatory compliance and reputation, which underscore the broader economic and regulatory benefits of adopting GRL practices.

Figure 10.3 shows the publications by country.

Figure 10.3 shows the distribution of publications on green reverse logistics (GRL) practices, and circular economy contributions underscores South Africa's prominent leadership role within Africa, representing 30% of the publications. This positions South Africa as a proactive leader in advancing sustainable practices on the continent, setting a significant example for other African nations to follow. The moderate engagement from India, Iran, and the United Kingdom, each contributing 7%, reflects a global interest in GRL and circular economies, while countries contributing 3% each demonstrate a broad international engagement in these critical sustainability areas. This diverse geographic spread highlights the universal relevance of GRL practices in addressing global sustainability challenges and indicates potential for substantial growth in research and implementation efforts, particularly in regions with lower current contributions. Africa's participation not only validates its commitment to sustainability but also positions the continent as a pivotal player in shaping future sustainable development strategies globally, leveraging regional strengths and fostering contextually relevant innovations.

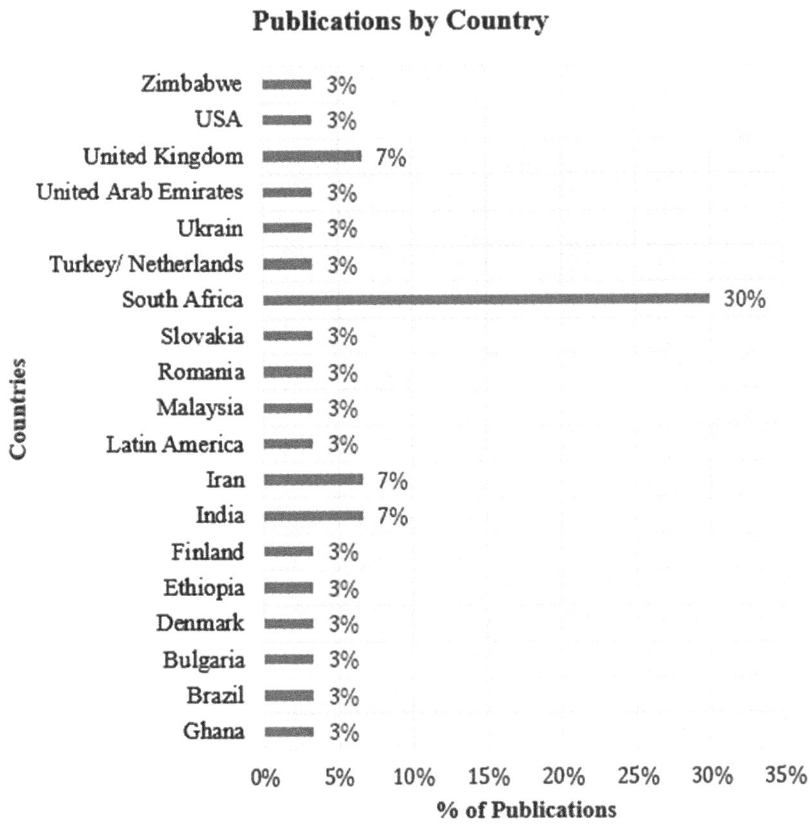

Fig. 10.3 Publications by country. (Source: Secondary data (2024))

According to Fig. 10.4, both Africa and the rest of the world frequently engage in practices such as recycling and material recovery, reverse logistics networks, remanufacturing and refurbishing, and stakeholder collaboration. However, the rest of the world exhibits more diversity with practices like product design for disassembly, environmental management systems, product take-back programs, and incentive programs, which are less reported in Africa.

According to Fig. 10.5, resource conservation and waste minimisation are the top circular economy contributions in both regions. These contributions are fundamental to the circular economy as they involve the

Fig. 10.4 GRL practices in Africa vs. rest of the world. (Source: Secondary data (2024))

minimisation of waste and the effective use of resources, which are crucial for sustainability. Resource conservation ensures that materials are used optimally and preserved for future use, while waste minimisation focuses on reducing the amount of waste generated, thus lowering environmental impact and improving the overall efficiency of production and consumption processes.

Additionally, the rest of the world places significant emphasis on contributions such as regulatory compliance and reputation, improved customer–supplier relations, and resilience to supply chain risks. Regulatory compliance and reputation involve adhering to environmental regulations and standards, which not only assists in avoiding fines but also improves the organisation's reputation among stakeholders, including consumers, investors, and regulators. Improved customer–supplier relations are vital for fostering trust and collaboration within the supply chain, leading to more efficient and effective operations. Resilience to supply chain risks addresses the ability to withstand and recover from disruptions, which is increasingly important in today's global and interconnected markets. These additional contributions showcase a broader approach to circular

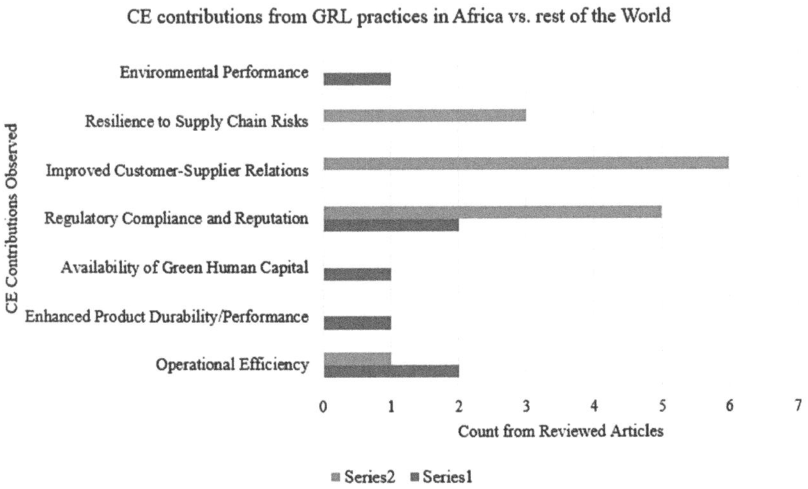

Fig. 10.5 CE contributions from GRL practices in Africa vs. rest of the world. (Source: Secondary data (2024))

economy practices as they incorporate elements of regulatory adherence, stakeholder engagement, and risk management, furthering the comprehensive sustainability agenda. By adopting such practices, African countries can benefit from a more holistic and integrated approach to circular economies, ensuring long-term environmental and economic resilience.

Discussion

The analysis of green reverse logistics (GRL) practices reveals that both Africa and other regions engage in foundational activities such as recycling and material recovery, reverse logistics networks, remanufacturing and refurbishing, and stakeholder collaboration. These practices are pivotal in promoting efficient resource use and reducing waste, which are critical for the development of circular economies. However, the broader range of GRL practices observed in other regions, including product design for disassembly, environmental management systems, product take-back programs, and incentive programs, highlights significant opportunities for Africa. By adopting these additional practices, African countries can further enhance their circular economy efforts and achieve more comprehensive sustainability outcomes.

The findings of this study correspond with the existing literature, which emphasises the importance of GRL practices in advancing circular economies. Prior research has consistently identified recycling and material recovery as central to circular economy frameworks (Sibanda et al., 2025). The broader adoption of practices such as Environmental Management Systems and Product Design for Disassembly in non-African regions is well-documented (Munuhwa & Hove-Sibanda, 2024). This study contributes to the literature by providing a detailed comparison between African practices and those prevalent in other parts of the world, thereby highlighting specific areas where African countries can expand their GRL initiatives and expand circular economies.

Theoretically, this research broadens understanding of how regional contexts influence the adoption and effectiveness of GRL practices. It provides a nuanced view of the intersection between local conditions and the implementation of circular economy strategies. Practically, the study offers a roadmap for African businesses and policymakers to adopt a wider array of GRL practices. The structured framework presented in the study, which includes practices like product design for disassembly, environmental management systems, and advanced technologies, serves as a practical guide for enhancing circular economy efforts in Africa. Additionally, the emphasis on stakeholder collaboration underscores the need for cooperative efforts among businesses, governments, and NGOs to create a cohesive and effective circular economy ecosystem. These insights are instrumental in driving policy formulation and strategic planning for African perspectives on sustainable development.

Implications

The research contributes to the existing body of knowledge in several significant ways: by comparing GRL practices and their outcomes between Africa and the rest of the world, it identifies specific practices that can be successfully implemented in the African context to enhance circular economies. The discussion provides a structured framework for African businesses and policymakers to adopt GRL practices, highlighting the benefits of practices such as product design for disassembly, environmental management systems, and advanced technologies. The comparative analysis sheds light on the differences and similarities in GRL practices and circular economy contributions between regions. This insight is crucial for understanding how local contexts influence the effectiveness of GRL strategies and for tailoring these strategies to fit African needs. Additionally, the chapter offers practical recommendations for African countries to adopt comprehensive GRL practices, fostering sustainable development and improved environmental outcomes. These recommendations serve as a guide for policymakers, businesses, and other stakeholders involved in circular economy initiatives. By emphasising the importance of stakeholder collaboration, the study encourages a cooperative approach to circular economy practices, advocating for partnerships between businesses, governments, and NGOs to create a cohesive and effective circular economy ecosystem.

Study Limitations

The study is susceptible to publication bias, potentially favouring studies from more developed regions with robust research infrastructures, which could skew the representation of certain countries or regions like Africa. Secondly, the accuracy and completeness of the publication databases used may vary, affecting the reliability and comprehensiveness of the study's findings. Lastly, the study may oversimplify the complexity of GRL practices and circular economy impacts by not fully accounting for regional contexts, cultural norms, and economic disparities, thereby limiting its ability to capture localised variations in implementation and effectiveness. These factors collectively emphasise the need for careful interpretation of the study's conclusions and identify areas for future research refinement. However, the studies that were included and reviewed in this research helped to bring insight on various GRL practices and their contribution towards the creation and development of the circular economy.

Conclusion

The study into green reverse logistics (GRL) practices and their potential to enhance circular economies in Africa has yielded several important insights. Both Africa and other regions frequently engage in GRL practices such as recycling and material recovery, reverse logistics networks, remanufacturing and refurbishing, and stakeholder collaboration. These practices are foundational to the development of circular economies, promoting efficient resource use and waste reduction. However, the rest of the world exhibits a broader range of GRL practices, including product design for disassembly, environmental management systems, product take-back programs, and incentive programs. These practices are less reported in Africa but offer significant potential for advancing circular economy goals on the continent. Resource conservation and waste minimisation emerge as the top circular economy contributions in both regions, underscoring their universal importance. Additionally, the rest of the world emphasises contributions such as regulatory compliance and reputation, improved customer–supplier relations, and resilience to supply chain risks. These additional contributions highlight a more comprehensive approach to circular economy practices, which Africa can benefit from by adopting similar strategies.

References

African Circular Economy Alliance (ACEA). (2023). *Annual report 2023: Advancing circular economy practices in Africa*. African Circular Economy Alliance. Retrieved June 20, 2024, from https://www.acea.org/annual-report-2023

Aithal, S., & Aithal, P. S. (2023). Importance of circular economy for resource optimisation in various industry sectors: A review-based opportunity analysis. *International Journal of Applied Engineering and Management Letters (IJAEML), 7*(2), 191–215. https://doi.org/10.5281/zenodo.8130738

Bag, S., & Gupta, S. (2019). Examining the effect of reverse logistics, green logistics, and green innovation practices on firm performance: A case of SMEs in Gauteng Province. In *Proceedings of the 31st annual conference of the Southern African Institute for Management Scientists (SAIMS)* (p. 312).

Barnes, J., & Morris, M. (2008). Staying alive in the global automotive industry: What can developing economies learn from South Africa about linking into global automotive value chains? *The European Journal of Development Research, 20*(1), 31–55.

Barney, J. (1991). Firm resources and sustained competitive advantage. *Journal of Management, 17*(1), 99–120.

Butt, A. S., Ali, I., & Govindan, K. (2023). The role of reverse logistics in a circular economy for achieving sustainable development goals: A multiple case study of retail firms. *Production Planning & Control*, 1–13.

Chartered Institute of Procurement & Supply (CIPS). (2021). *Sustainable and ethical procurement: Best practices and emerging trends*. Chartered Institute of Procurement & Supply. Retrieved June 28, 2024, from https://www.cips.org/sustainable-and-ethical-procurement

Coca-Cola Beverages Africa. (2020). *Coca-Cola Beverages Africa partners with local governments to boost plastic recycling efforts*. Coca-Cola Beverages Africa. Retrieved July 5, 2024, from https://www.coca-cola.com/za/en/media-center/coca-cola-takes-bold-actions-on-plastic-waste

Correa-Vaca, A. M., Miralles, C., Márquez-Sánchez, F., & Vergara-Romero, A. (2024). Circular economy based on reverse logistics: A systematic literature review. *Kurdish Studies, 12*(1), 3022–3037.

Derse, O. (2024). Prioritising solutions of green reverse logistics barriers with Fuzzy DEMATEL–FUCOM–SWARA methods. *Ecological Indicators, 165*, 112198.

Elkington, J. (1994). Towards the sustainable corporation: Win-win-win business strategies for sustainable development. *California Management Review, 36*(2), 90–100.

Ellen MacArthur Foundation. (2015). *Growth within: A circular economy vision for a competitive Europe*. Ellen MacArthur Foundation. Retrieved June 18, 2024, from https://ellenmacarthurfoundation.org/growth-within-a-circular-economy-vision-for-a-competitive-europe

Epoh, L. R., & Mafini, C. (2018). Green supply chain management in small and medium enterprises: Further empirical thoughts from South Africa. *Journal of Transport and Supply Chain Management, 12*(1), 1–12.

European Commission. (2020). *A new circular economy action plan: For a cleaner and more competitive Europe*. European Commission. Retrieved July 4, 2024, from https://ec.europa.eu/environment/strategy/circular-economy-action-plan_en

Fernando, Y., Shaharudin, M. S., & Abideen, A. Z. (2022). Circular economy-based reverse logistics: Dynamic interplay between sustainable resource commitment and financial performance. *European Journal of Management and Business Economics, 32*(1), 91–112.

Gautam, M., & Agrawal, M. (2021). Greenhouse gas emissions from municipal solid waste management: A review of global scenario. In *Carbon footprint case studies: Municipal solid waste management, sustainable road transport and carbon sequestration* (pp. 123–160). https://doi.org/10.1007/978-981-15-9577-6_5

Geissdoerfer, M., Savaget, P., Bocken, N. M. P., & Hultink, E. J. (2017). The circular economy—A new sustainability paradigm? *Journal of Cleaner Production, 143*, 757–768.

Gharibi, K., & Abdollahzadeh, S. (2021). A mixed-integer linear programming approach for circular economy-led closed-loop supply chains in green reverse logistics network design under uncertainty. *Journal of Enterprise Information Management*. https://doi.org/10.1108/JEIM-11-2020-0472

Gobena, W. T., & Azene, D. K. (2022). Adoption of reverse logistics to waste plastic recycling: Investigation of critical success factors in Ethiopia. *International Journal of Sustainable Society, 14*(1), 43–76.

Godfrey, L., & Oelofse, S. (2017). Historical review of waste management and recycling in South Africa. *Resources, 6*(4), 57.

Govindan, K., Nosrati-Abarghooee, S., Nasiri, M. M., & Jolai, F. (2022). Green reverse logistics network design for medical waste management: A circular economy transition through case approach. *Journal of Environmental Management, 322*, 115888.

Helfat, C. E., Kaul, A., Ketchen, D. J., Jr., Barney, J. B., Chatain, O., & Singh, H. (2023). Renewing the resource-based view: New contexts, new concepts, and new methods. *Strategic Management Journal, 44*(6), 1357–1390.

Jayarathna, C. P., Agdas, D., & Dawes, L. (2024). Viability of sustainable logistics practices enabling circular economy: A system dynamics approach. *Business Strategy and the Environment, 33*(4), 3422–3439.

Khan, R., & Mihaisi, S. M. K. A. (2023). Promoting circular economy model through SMEs' growth: A focus on African nations. *Journal of Production Operations Management and Economics.* https://doi.org/10.55529/jpome.36.40.52

Khan, S. A. R., Dong, Q. L., Zhang, Y. U., & Khan, S. S. (2017). Research on decision-making of green reverse logistics in enterprises: A case study on electronic products manufacturers from the perspective of South Africa. In *Proceedings of the 2017 2nd International Conference on Advances in Management Engineering and Information Technology (AMEIT)* (Vol. 2017, pp. 11–18).

Koech, M. K., Munene, K. J., Kinoti, M. K., & Situma, D. K. (2023). Circular economy policies and innovations in Africa: Pillars for achieving sustainable development. In S. K. Ghosh & S. K. Ghosh (Eds.), *Circular economy adoption* (pp. 55–76). Springer.

Makaleng, M. S. M., & Hove-Sibanda, P. (2022). Reverse logistics strategies and their effect on the competitiveness of fast-moving consumer goods firms in South Africa. *Logistics, 6*(3), 56.

Mallick, P. K., Salling, K. B., Pigosso, D. C. A., & McAloone, T. C. (2023). Towards a circular economy: Development of a support tool for designing reverse logistics systems. *Journal of Environmental Management, 351*, 119819.

Mayanti, B., & Helo, P. (2024). Circular economy through waste reverse logistics under extended producer responsibility in Finland. *Waste Management & Research, 42*(1), 59–73.

Mingst, K. A., Karns, M. P., & Lyon, A. J. (2022). *The United Nations in the 21st century.* Routledge. https://doi.org/10.4324/9781003038269

Mugoni, E., Nyagadza, B., & Hove, P. K. (2023). Green reverse logistics technology impacts on agricultural entrepreneurial marketing firms' operational efficiency and sustainable competitive advantage. *Sustainable Technology and Entrepreneurship, 2*(2), 100034.

Munuhwa, S., & Hove-Sibanda, P. (2024). Exploring sustainable food supply chain management practices to enhance food security. *Journal of Transport and Supply Chain Management, 18*(0), a1064. https://doi.org/10.4102/jtscm.v18i0.1064

Ofori, P. (2023). Circular economy practices in third world nations: Challenges and implications for environmental sustainability. *Environment, Development and Sustainability*, 1–55.

Ogunseye, N. O., Odufuwa, B. O., Owolabi, M. A., & Tella, I. O. (2020). Solid waste recycling: A severe problem in Nigeria. *The Journal of Solid Waste Technology and Management, 46*(3), 442–454.

Ozdemir, S., de Arroyabe, J. C. F., Sena, V., & Gupta, S. (2023). Stakeholder diversity and collaborative innovation: Integrating the resource-based view with stakeholder theory. *Journal of Business Research, 164*, 113955.

Page, M. J., McKenzie, J. E., Bossuyt, P. M., Boutron, I., Hoffmann, T. C., Mulrow, C. D., et al. (2021). The PRISMA 2020 statement: An updated guideline for reporting systematic reviews. *BMJ, 372.*

Ritter, M., Schilling, H., & Brüggemann, H. (2024). Towards achieving the sustainable development goals: A collaborative action plan leveraging the circular economy potentials. *Gr Interakt Org, 55*, 175–187. https://doi.org/10.1007/s11612-024-00733-9

Scrioșteanu, A., & Criveanu, M. M. (2024). Green and reserve logistics of packaging and plastic packaging waste under the conditions of circular economy at the level of the European Union member states. *Energies, 17*(12), 3015.

Sibanda, V., Mhlanga, J., & Munuhwa, S. (2025). Circular Economy Practices in Supply Chain Management. In A. Asa & J. Nautwima (Eds.), *Eco-Logistics and Sustainable Supply Chain Innovations* (pp. 13-38). IGI Global Scientific Publishing. https://doi.org/10.4018/979-8-3693-6175-7.ch002

Srivastava, A. K., Dixit, S., & Srivastava, A. A. (2022). Criticism of triple bottom line: TBL (with special reference to sustainability). *Corporate Reputation Review*, 1–12.

The World Bank. (2018). *What a waste 2.0: A global snapshot of solid waste management to 2050.* The World Bank. https://doi.org/10.1596/978-1-4648-1329-0

United Nations Environment Programme (UNEP). (2020). *Sustainable public procurement: A global review 2020.* United Nations Environment Programme. Retrieved June 25, 2024, from https://www.unep.org/resources/report/sustainable-public-procurement-global-review-2020

WEEE Centre. (2021). *Waste Electrical and Electronic Equipment Centre.* WEEE Centre. Retrieved June 23, 2024, from https://weeecentre.com/about-us/

World Economic Forum. (2021). Why the circular economy is a billion-dollar opportunity for Africa. *World Economic Forum.* Retrieved June 25, 2024, from https://www.weforum.org

CHAPTER 11

Artificial Intelligence-Driven Green Marketing Strategies for a Circular Economy

Martin Muduva and Fungai Jacqueline Kiwa

Introduction

Artificial intelligence (AI) is fundamentally transforming industries worldwide, with a particularly profound impact on green marketing strategies. By harnessing the power of AI, businesses can optimize resource utilization through the analysis of large datasets, effectively predicting demand and managing resources, which leads to significant reductions in waste and energy consumption across manufacturing and supply chains (Akira, 2023). AI-driven predictive analytics empower companies to minimize excess inventory, thereby lessening their environmental footprint and

M. Muduva (✉)
Department of Information and Marketing Science, Midlands State University, Gweru, Zimbabwe
e-mail: muduvam@staff.msu.ac.zw

F. J. Kiwa
Department of ICT and Electronics, School of Engineering Science and Technology, Chinhoyi University of Technology, Chinhoyi, Zimbabwe
e-mail: jkiwa@cut.ac.zw

A. Muposhi, B. Nyagadza (eds.), *Sustainable Green Marketing Strategies for a Circular Economy*, Palgrave Studies of Marketing in Emerging Economies,
https://doi.org/10.1007/978-3-031-81627-7_11

enhancing their sustainability initiatives (Wang et al., 2020). Advanced AI tools such as machine learning and natural language processing (NLP) allow for precise segmentation of consumers based on their eco-conscious behaviors, enabling the development of highly targeted and personalized green marketing campaigns (Bhardwaj et al., 2020). For instance, retailers can tailor their marketing strategies to resonate with consumers interested in sustainable products, thereby boosting engagement and promoting eco-friendly choices (Liu et al., 2020). AI significantly enhances supply chain transparency, enabling companies to monitor and optimize operations while providing real-time data on product sourcing and sustainability efforts, which supports effective green marketing strategies (Choi & Kim, 2021). AI-powered platforms facilitate the tracking of raw materials' sustainability, allowing businesses to authentically promote their commitment to ethical sourcing in their marketing communications (Tseng et al., 2020). AI's capacity to predict consumer preferences and behaviors regarding sustainability allows companies to refine their marketing strategies, encouraging eco-friendly purchasing decisions (Kaplan et al., 2021). With its advanced capabilities for automation, personalization, and data analysis, AI equips businesses to design marketing campaigns that are not only more efficient but also more sustainable.

This transition is vital as companies pivot towards circular economy models, which emphasize resource efficiency, waste reduction, and the extension of product lifecycles (Jiang et al., 2022). Unlike the conventional linear model characterized by a "take, make, waste" cycle, the circular economy aims to reuse, recycle, and regenerate resources, thereby aligning with contemporary sustainability objectives (Ellen MacArthur Foundation, 2019). The circular economy model underscores the importance of minimizing waste and extending product lifecycles, fostering innovation and resource efficiency (Geissdoerfer et al., 2018). Recent research indicates that embracing circular economy principles can not only mitigate environmental impact but also enhance economic resilience and competitiveness (Witjes & Lozano, 2016). The integration of technologies such as artificial intelligence and the Internet of Things plays a crucial role in facilitating circular economy practices by improving resource tracking and management (Bocken et al., 2021). As businesses increasingly recognize the necessity of sustainable practices, the shift towards a circular economy signifies a major transformation in industry operations and value delivery (Murray et al., 2017). Amid growing environmental awareness, businesses face mounting pressure to meet sustainability demands while maintaining competitiveness. AI addresses these needs by refining marketing strategies, enabling businesses to anticipate consumer behavior, and optimizing resource use (Agrawal et al., 2022). Machine learning-driven predictive models afford companies the ability to tailor messages according to individual consumer preferences and eco-conscious behaviors, enhancing engagement and fostering customer loyalty (Patel & Singh, 2023). This targeted approach ensures that marketing efforts are not only effective but also aligned with sustainability goals, consequently reducing resource waste throughout the product lifecycle (Chen & Chang, 2021).

AI's ability to process substantial volumes of data in real-time allows marketers to adapt their strategies as consumer trends and environmental considerations evolve (Dwivedi et al., 2021). AI systems can analyze consumer sentiments regarding sustainability across social media platforms, yielding insights that inform more relevant marketing campaigns (Nzomo et al., 2023). This level of responsiveness is especially crucial in fast-changing markets, particularly in regions like Africa, where resource scarcity remains a pressing challenge (Nwankwo & Mbachu, 2022). The application of AI in sustainable marketing holds considerable promise in emerging markets, where businesses frequently contend with resource and technological limitations (Mishra & Sung, 2022). There is a notable lack of empirical evidence regarding the real-world application of AI in these resource-constrained environments, indicating a significant gap in current research (Adeola & Ezenwafor, 2021). While African start-ups illustrate the potential of AI to drive eco-innovation in sectors such as agribusiness and energy (Nwankwo & Mbachu, 2022), more investigation is needed into how AI technologies can be tailored to meet the specific challenges faced by these markets, including disparities in technological infrastructure and varying levels of consumer awareness (Mukonza et al., 2021). This study aims to address these gaps by providing insights into how AI can foster sustainable practices within African markets. It offers a roadmap for businesses and policymakers to leverage AI in achieving sustainability objectives, emphasizing resource optimization, supply chain management, and consumer engagement (Geissdoerfer et al., 2020; Lacy & Rutqvist, 2021). By exploring real-world applications of AI in African markets, this research contributes to the expanding body of knowledge on AI-driven sustainability, presenting practical solutions for promoting economic growth and environmental stewardship in emerging economies.

This chapter aims to explore the potential of AI-driven green marketing strategies in promoting sustainability through the lens of the circular economy, with a specific focus on African markets. The chapter addresses the following objectives:

- Analyze how AI technologies can optimize resource management and reduce waste in marketing efforts.
- Examine AI's role in fostering sustainable consumer behaviors through targeted, data-driven campaigns.
- Analyze case studies from African markets to illustrate the tangible benefits of integrating AI into green marketing strategies, particularly in terms of consumer engagement and resource efficiency.

These objectives will provide insights for businesses and policymakers on how to effectively leverage AI for sustainability and gain a competitive edge in the evolving green market landscape.

Theoretical Foundations

The theoretical development of this chapter draws from three core areas: disruptive innovation theory, diffusion of innovations theory, and the social cognitive theory. These fields are crucial in understanding how AI can be leveraged to optimize resource management and foster sustainable consumer behaviors through data-driven green marketing strategies. Integrating these disciplines, businesses can not only engage consumers with eco-conscious campaigns but also reduce waste in their marketing efforts. This approach aligns with broader sustainability goals, ensuring that AI-driven marketing promotes both environmental stewardship and operational efficiency.

Disruptive Innovation Theory

Disruptive innovation theory posits that innovations can transform industries by introducing new products or services that initially serve a niche market but eventually displace established competitors (Christensen et al., 2018). In the context of green marketing, this theory elucidates how advancements in artificial intelligence (AI) can disrupt traditional marketing practices and consumer behavior by making sustainable products more accessible and appealing (Bocken et al., 2021). The rise of AI in marketing enables companies to leverage data analytics to identify and target emerging consumer trends toward eco-friendly products, thus creating new market opportunities (Witjes & Lozano, 2016). Utilizing sophisticated algorithms, businesses can analyze vast datasets from various sources, such as online behaviors, social media interactions, and purchasing history, to uncover insights into consumers' growing preferences for sustainability (Zhang et al., 2023). This ability to tap into real-time data allows companies to respond quickly to changing consumer attitudes, making it possible to craft targeted marketing strategies that align with eco-conscious consumer values. AI-driven platforms can introduce innovative features, such as personalized recommendations and tailored marketing campaigns that resonate with consumers' values, thus attracting a broader audience to green products (Huang et al., 2021). For example, AI can track consumer interactions and feedback to refine product offerings and promotional strategies, creating a feedback loop that enhances engagement. This level of personalization is crucial, as research indicates that consumers are more likely to purchase products that they feel are aligned with their personal beliefs and values (Kumar et al., 2023).

The disruption is further amplified by AI's ability to streamline the shopping experience, offering intelligent suggestions, personalized promotions, and easy access to eco-friendly alternatives, which encourage consumers to adopt sustainable choices (Kim & Park, 2023). With AI tools, businesses can automate the marketing process, optimizing it for various segments and ensuring that the messaging is relevant and timely. This reduces friction in the buying process, allowing consumers to make more informed decisions about their purchases.

Minimizing barriers to sustainable consumption and enhancing consumer confidence, AI transforms how individuals perceive and interact with green products (Kumar et al., 2023). The technology not only provides information but also instills a sense of trust, as AI can offer transparency about product origins, sustainability practices, and overall environmental impact. When consumers are better informed, they are more likely to engage with and invest in sustainable products. The integration of AI into marketing strategies aligns with the principles of disruptive innovation theory, allowing businesses to target consumer segments that are increasingly inclined towards sustainability (Patel & Singh, 2023). By fostering a culture of eco-consciousness and promoting the demand for green products, AI not only disrupts traditional marketing but also supports the objectives of the circular economy, ultimately driving a more sustainable marketplace (Geissdoerfer et al., 2020). As the market landscape evolves, companies that effectively leverage AI in their green marketing strategies are likely to gain a competitive advantage, positioning themselves as leaders in sustainability and innovation.

Diffusion of Innovations Theory

The diffusion of innovations (DOI) theory explains the process by which new ideas, products, or behaviors spread through a population. The theory identifies several adopter categories: innovators, early adopters, early majority, late majority, and laggards (Rogers, 2003). In the context of green marketing, the DOI theory helps marketers understand how AI can be leveraged to promote sustainable products and drive adoption of eco-friendly practices across these different consumer groups (Huang et al., 2021; Evans & Cooper, 2022). Early adopters, who are often trendsetters, play a crucial role in influencing the behaviors of others (Patel & Singh, 2023; Kim & Park, 2023). AI enables businesses to identify these key influencers by analyzing social network interactions, purchasing behaviors and even geolocation data to understand who is driving green purchasing trends (Zhang et al., 2023). Once early adopters are identified, AI-driven marketing platforms can focus on nurturing this group with targeted content that highlights the environmental and social benefits of green products (Huang et al., 2021). Fostering strong connections with early adopters, companies can accelerate the diffusion process, as this group will naturally promote the products through their networks, encouraging the early and late majority to follow suit (Evans & Cooper, 2022; Patel & Singh, 2023). AI also facilitates real-time feedback loops, allowing businesses to adjust their marketing strategies to the shifting behaviors and preferences of these groups, thus enhancing the adoption curve of sustainable innovations (Zhang et al., 2023).

AI algorithms can track the diffusion of green products over time, offering insights into the most effective points of entry for marketing eco-friendly products across different consumer segments (Kim & Park, 2023). This dynamic approach ensures that businesses are continuously optimizing their marketing efforts to accelerate the spread of sustainable behaviors in line with the principles of the circular economy (Evans & Cooper, 2022; Patel & Singh, 2023). Through leveraging AI technologies in alignment with DOI theory, businesses can accelerate the spread of green innovations, ensuring that eco-friendly products reach wider audiences more efficiently (Zhang et al., 2023; Huang et al., 2021). AI's ability to identify early adopters and track behavioral patterns allows companies to strategically enhance the adoption of sustainable products, thereby supporting circular economy goals (Evans & Cooper, 2022).

Social Cognitive Theory

Social cognitive theory (SCT), developed by Albert Bandura, emphasizes that behavior is learned through observation, imitation, and modeling (Bandura, 1986). People acquire new behaviors by watching others, particularly influential figures or peers, and the social context in which behaviors occur plays a significant role in decision-making (Huang et al., 2021). In the realm of AI-driven green marketing, social cognitive theory offers insights into how AI can be used to influence consumer behavior toward sustainable practices (Zhang et al., 2023). AI can analyze social networks, peer influence, and online communities to understand how eco-friendly behaviors are modeled and spread within these networks (Patel & Singh, 2023). AI can identify influential social media figures or community leaders who advocate for sustainability and marketers can collaborate with these influencers to promote green products (Huang et al., 2021). By highlighting the positive social impact of adopting eco-friendly behaviors, businesses can encourage consumers to imitate these actions, especially when such behaviors are perceived as being endorsed by respected peers or role models (Zhang et al., 2023). AI also allows companies to monitor the success of these social campaigns in real time, ensuring that messages are adapted to remain relevant and impactful as consumer sentiment shifts (Patel & Singh, 2023). AI-powered recommendation systems can be designed to reinforce positive behaviors by continuously promoting sustainable products to consumers who have shown interest in eco-friendly practices (Huang et al., 2021). These recommendations create a feedback loop that further strengthens consumer commitment to sustainability by consistently reinforcing the social value of green behavior (Zhang et al., 2023). AI enhances social cognitive theory's application in green marketing by enabling marketers to understand and influence consumer behavior through the modeling of sustainable practices (Patel & Singh, 2023). Promoting eco-friendly behaviors that are seen as socially desirable and endorsed by influencers, AI-driven marketing strategies encourage greater consumer engagement with green products, thus supporting the shift toward a circular economy (Huang et al., 2021).

Integration of AI and Marketing for Sustainability

The integration of AI into marketing significantly enhances the precision and personalization of campaigns, allowing businesses to create highly targeted messages that resonate with consumers' environmental values (Sun et al., 2020). Leveraging machine learning algorithms, marketers can effectively segment consumers based on their eco-conscious preferences, tailoring messages that highlight sustainable products and practices (Belanche et al., 2021). This personalized approach not only increases consumer engagement but also fosters deeper connections with eco-conscious brands, driving brand loyalty in the long term (Mishra et al., 2020). AI's ability to analyze vast amounts of consumer data enables it to predict consumer behavior and preferences with a high degree of accuracy, which helps businesses allocate resources more effectively (Yadav & Gupta, 2021). Predictive analytics also allow companies to stay competitive by identifying emerging trends in sustainability, ensuring that their marketing strategies proactively address consumer demands for green products (Obaidi et al., 2022). This foresight improves the efficiency of marketing campaigns and ensures that businesses remain relevant in an increasingly eco-conscious market (Yadav & Gupta, 2021).

AI-driven marketing plays a vital role in enhancing environmental sustainability by optimizing resource management within marketing operations (Huang et al., 2021). For example, AI can evaluate the environmental impact of various marketing channels and recommend the eco-friendliest options to reduce waste and energy consumption (Okoro & Nwachukwu, 2022). In addition, AI-driven tools can help optimize supply chain processes, minimizing the carbon footprint of marketing activities by streamlining production and distribution channels (Mwenda et al., 2023). These optimizations align marketing strategies with broader corporate sustainability goals, enhancing a company's reputation for environmental responsibility (Moyo & Ndlovu, 2023).

Enhancing Customer Experiences Through AI

AI-powered tools such as chatbots and recommendation engines are transforming the customer experience by providing personalized interactions that promote green products and services (Dwivedi et al., 2021). These AI-driven tools enable businesses to guide consumers toward eco-friendly purchasing options, helping them make more environmentally responsible decisions (Sun et al., 2020). Analyzing consumer preferences and behavior in real time, AI enhances the relevance of product recommendations, making it easier for customers to choose sustainable products (Mishra & Sung, 2022). AI allows businesses to continuously monitor and analyze customer feedback, enabling them to refine their marketing strategies and improve the delivery of sustainable products (Luo & Wilson, 2021). Sentiment analysis tools can track customer opinions and attitudes toward green products, helping marketers adjust their messaging to better meet consumer expectations (Obaidi et al., 2022). This feedback loop not only strengthens the relationship between consumers and eco-friendly brands but also increases long-term brand loyalty based on shared environmental values (Belanche et al., 2021). AI also helps businesses personalize the shopping experience for each customer, ensuring that eco-conscious consumers receive tailored recommendations that align with their sustainability goals (Yadav & Gupta, 2021). Improving the convenience and accessibility of sustainable products, AI fosters greater engagement with green brands, ultimately encouraging more responsible consumer behavior (Dwivedi et al., 2021).

AI-Driven Strategies in African Markets

The application of AI in African markets presents unique opportunities and challenges due to the diversity of consumer behavior, technological adoption, and environmental issues (Emon & Khan, 2024). AI enables businesses to segment consumer groups based on local preferences and environmental concerns, allowing marketers to develop highly effective campaigns that resonate with specific regions (Emon & Khan, 2024). This level of customization is essential in African markets, where cultural and environmental factors significantly influence consumer behavior (Mensah et al., 2021). The uneven distribution of technological infrastructure across Africa requires adaptable AI strategies to address varying levels of technological maturity in different regions (Chimuka, 2022). Countries like South Africa and Kenya have relatively advanced technological infrastructure, making them ideal markets for AI-driven green marketing (Mensah et al., 2021). In contrast, other countries may need more foundational development before AI can be fully integrated into marketing operations, requiring businesses to tailor their approaches accordingly (Nzomo et al., 2023). In terms of technological disparities, African markets face environmental challenges such as resource scarcity and the impacts of climate change (Mensah et al., 2021). AI can help address these challenges by optimizing resource use in marketing activities and improving the efficiency of sustainability initiatives (Moyo & Ndlovu, 2023). AI-driven tools can analyze environmental data to identify areas where businesses can reduce waste, conserve energy, and adopt more sustainable practices (Mwenda et al., 2023). These strategies not only support local economies but also promote sustainability at a regional and global level (Achebe, 2020).

AI for Promoting a Circular Economy

A circular economy emphasizes minimizing waste and reusing materials to create closed-loop systems that reduce environmental impact (Okoro & Nwachukwu, 2022). AI plays a crucial role in promoting this transition by enabling businesses to develop personalized marketing campaigns that highlight the benefits of recycling, reusing, and reducing consumption (Mwenda et al., 2023). AI-driven tools can analyze consumer behavior to identify individuals who are more likely to adopt circular economy practices, allowing businesses to target them with tailored messages that promote sustainable products (Achebe, 2020). AI can recommend products made from recycled materials or second-hand goods based on a consumer's purchasing history and environmental values (Obaidi et al., 2022). This helps create a demand for circular economy products, fostering a shift toward more sustainable consumption patterns (Emon & Khan, 2024). AI-driven supply chain optimization can reduce waste and promote the use of sustainable materials, further aligning marketing efforts with the principles of a circular economy (Belanche et al., 2021). AI can also help businesses track the lifecycle of their products, ensuring that materials are reused or recycled at the end of their lifecycle (Huang et al., 2021). This enables companies to reduce waste and improve the efficiency of their operations, supporting the goals of a circular economy (Emon & Khan, 2024). Integrating AI into their sustainability strategies, businesses can not only enhance the effectiveness of their marketing but also contribute to long-term environmental goals (Sun et al., 2020).

Empirical Review

The intersection of artificial intelligence (AI), green marketing, and sustainability has become an area of growing academic interest and practical application. This section critically reviews the relevant literature, exploring the application of AI in green marketing and its potential to support a circular economy. It also highlights gaps in the research and identifies areas where this study contributes to the existing body of knowledge. Recent literature extensively documents the transformative impact of AI on marketing strategies, particularly through machine learning and predictive analytics. Emon and Khan (2024) emphasize that AI enables marketers to develop more targeted and personalized campaigns by analyzing large datasets, offering deeper insights into consumer behavior and preferences. These capabilities enhance the precision of marketing efforts, making campaigns more effective and resource-efficient (Kumar & Prakash, 2020). Automating routine tasks, AI improves customer segmentation, optimizes marketing spend, and enhances operational efficiency, as highlighted by Emon and Khan (2024). Gupta and Sharma (2023) further argue that AI-driven marketing significantly enhances customer engagement and retention, providing businesses with a competitive edge in fast-evolving markets.

Green marketing, a key component of sustainability efforts, promotes products and practices that minimize environmental impact. Kumar and Prakash (2020) outline the core principles of green marketing, including reducing environmental harm, promoting sustainable consumption, and enhancing corporate social responsibility. These principles are essential for aligning marketing strategies with sustainability goals and consumer expectations. Chen and Chang (2021) discuss the drivers and barriers to adopting green marketing, noting the importance of consumer awareness and corporate commitment to sustainability. The role of consumer education is critical, as Agyeman et al. (2021) suggest that informed consumers are more likely to support green products, thus fostering the success of sustainable marketing strategies. The convergence of AI and green marketing is an emerging area of study that offers significant potential. Patel and Singh (2023) explore how AI supports green marketing by analyzing consumer preferences for sustainable products and optimizing resource use in marketing campaigns. Their findings indicate that AI can drive both economic and environmental benefits by improving the efficiency of marketing operations and reducing waste (Huang et al., 2021). AI's ability to identify new market opportunities for eco-friendly products, as demonstrated by Emon and Khan (2024), shows that AI can facilitate business growth while promoting environmental stewardship. This emerging convergence offers a path toward sustainable marketing practices that are both economically viable and environmentally responsible.

The concept of a circular economy, which focuses on minimizing waste and maximizing resource efficiency, is closely aligned with sustainability goals. According to Geissdoerfer et al. (2020), the circular economy incorporates principles such as recycling, reusing, and remanufacturing, which are essential for reducing environmental impacts. Emon and Khan (2024) argue that digital technologies, including AI, can facilitate the transition to a circular economy by optimizing resource flows. Lacy and Rutqvist (2021) further illustrate how AI can enhance supply chain management, product lifecycle analysis, and waste reduction, contributing to sustainable development. Evans and Cooper (2022) provide evidence that AI-driven predictive analytics can improve recycling processes, leading to higher rates of material recovery and reuse, further supporting the circular economy. Despite the promising intersections of AI, green marketing, and circular economy strategies, several research gaps remain. While AI adoption has been shown to enhance marketing efficiency and sustainability, there is limited evidence of its practical application in resource-constrained environments, particularly in Africa (Adeola & Ezenwafor, 2021). Studies focusing on developed economies often fail to account for the unique challenges faced by emerging markets, such as technological infrastructure disparities and varying levels of consumer awareness regarding sustainability (Mukonza et al., 2021). Moreover, although AI has the potential to support eco-innovation and resource efficiency in African markets, further research is needed to understand how these technologies can be effectively scaled and integrated into local business models (Nwankwo & Mbachu, 2022).

Case Studies and Applications in African Markets

While much of the existing literature focuses on developed markets, a growing body of research examines AI and green marketing in emerging economies, particularly in Africa. Adeola and Ezenwafor (2021) suggest that AI adoption by African businesses can significantly enhance the efficiency of green marketing initiatives, even in resource-constrained environments. Their findings emphasize the potential for AI to drive both sustainability and economic growth in Africa. Hinson et al. (2021) and Mukonza et al. (2021) identify several challenges and opportunities in implementing sustainable marketing strategies across African markets, including the need for increased technological infrastructure and consumer education. Nwankwo and Mbachu (2022) explore how Nigerian startups leverage AI to optimize resource allocation and target environmentally conscious consumers. Their research demonstrates that AI can contribute to both business growth and sustainability, highlighting its transformative potential in the African context. Similarly, Okolo-obasi and Uduji (2023) examine AI's role in Ghanaian agribusinesses, showing that AI-driven predictive analytics significantly improve resource efficiency and crop yields, contributing to sustainable agricultural practices. These case studies underscore the role of AI in promoting green marketing and sustainability in Africa, where environmental challenges and resource constraints often demand innovative solutions (Mwangi, 2024).

In Kenya's retail sector, AI-enabled customer segmentation and personalized marketing have significantly boosted the adoption of sustainable products, as noted by Mwangi (2024). AI's ability to analyze consumer data and provide tailored recommendations for eco-friendly products demonstrates its potential to drive responsible consumption (Chimuka, 2022). In South Africa, Adetunji and Oke (2020) find that AI-driven green marketing in the automotive industry enhances supply chain efficiency and customer engagement, leading to better sustainability outcomes. These case studies collectively illustrate how African businesses are leveraging AI to drive green marketing initiatives, fostering both environmental sustainability and economic growth (Emon & Khan, 2024).

Gaps and Contributions

Despite the progress made in understanding AI's role in green marketing and sustainability, significant research gaps remain, particularly concerning its application in emerging markets. Current studies indicate that while AI adoption enhances marketing efficiency and sustainability practices, there is a lack of real-world evidence from resource-constrained environments (Adeola & Ezenwafor, 2021). African startups have demonstrated the potential for AI to drive eco-innovation and resource efficiency, especially in sectors like agribusiness and energy (Nwankwo & Mbachu, 2022). More research is needed to explore how AI can be effectively tailored to address the unique challenges of these markets, such as technological infrastructure disparities and varying levels of consumer awareness (Mukonza et al., 2021). This study seeks to fill these gaps by offering insights into how AI can drive sustainable practices in African markets. It provides a roadmap for businesses and policymakers to leverage AI in achieving sustainability goals, with a focus on resource optimization, supply chain management and consumer engagement (Geissdoerfer et al., 2020; Lacy & Rutqvist, 2021). Examining real-world applications of AI in African markets, this study contributes to the growing body of knowledge on AI-driven sustainability, offering practical solutions for fostering economic growth and environmental stewardship in emerging economies.

Research Methodology

This section outlines the methodological approach employed in the study, which examines AI-driven green marketing strategies aimed at promoting a circular economy in African markets. The research investigates how businesses, particularly startups in Africa, leverage artificial intelligence to enhance resource efficiency, minimize waste, and encourage sustainable consumer behavior. The methodology is designed to ensure comprehensive, credible, and contextually relevant insights into the application of AI in promoting sustainability through green marketing practices. The study employs a qualitative approach, integrating case studies, interviews, and document analysis to uncover key insights. This approach allows for the collection of in-depth, nuanced data that can offer a deep understanding of the intersections between AI, green marketing, and sustainability in emerging markets. The selection of participants is purposive, with the goal of including organizations that are actively engaged in both AI adoption and green marketing initiatives.

Research Context

This study is conducted in the context of AI-driven green marketing strategies within Zimbabwe, specifically focusing on businesses in Harare. As the capital and economic center, Harare hosts a wide range of industries, making it an ideal location to investigate the use of AI technologies to enhance resource efficiency, reduce waste, and foster sustainable consumer behavior. The environmental, economic, and technological landscape in

Harare provides a fertile ground for examining the practical applications of AI in green marketing, particularly in businesses that are transitioning toward more sustainable practices.

Target Population

The target population for this study consists of businesses in Harare that are engaged in green marketing initiatives and are actively leveraging AI technologies. These businesses represent various sectors, including manufacturing, retail, and services, all with a vested interest in sustainability and resource management. Policymakers, industry experts, and key stakeholders from relevant organizations are also included to offer additional insights into the role of AI in driving green marketing and sustainability in Zimbabwe.

Sample Size

The sample size for this study has been carefully selected to ensure that it remains manageable while providing diverse perspectives. The study focuses on a sample of 15 to 20 businesses in Harare, each varying in size, industry, and level of AI adoption. This sample size allows for an in-depth exploration of the experiences of these businesses, providing a rich dataset while ensuring that the study remains feasible within the given timeframe and resources. The sample is designed to capture both large enterprises and small to medium-sized enterprises (SMEs), ensuring a representative cross-section of businesses operating in Harare's dynamic market.

Sample Selection

The sample selection process involved purposive sampling, targeting businesses in Harare that are known for their innovative use of AI in marketing and their commitment to sustainability. This selection criterion ensures that the study focuses on cases where AI-driven green marketing strategies are actively being implemented. Concentrating on Harare, the study can provide more nuanced insights into the specific challenges and opportunities businesses face in a localized context. Industry experts and policymakers were selected based on their expertise and involvement in promoting sustainability and AI adoption in Zimbabwe.

Credibility Ad Trustworthiness

To ensure analytical rigor, the study employs a qualitative methodology, focusing on collecting and analyzing detailed, context-rich data. This approach enables a thorough exploration of the real-world experiences and strategies employed by businesses in Harare. Data triangulation is achieved by using multiple data sources, such as interviews, case studies, and document analysis, to enhance the credibility of the findings. The combination of in-depth interviews with key stakeholders and case studies of AI applications in green marketing ensures that the research captures the complexity of the topic while remaining focused on the local context.

Data Collection

Data collection involved semi-structured interviews with business leaders, marketing professionals, and industry experts in Harare. These interviews were designed to gather detailed narratives on the implementation and impact of AI-driven green marketing strategies within the Zimbabwean market. Observational data and document analysis of company reports, marketing materials and sustainability statements were used to triangulate findings. This mixed-methods approach ensures that the research provides a comprehensive understanding of how businesses in Harare are using AI to achieve sustainability goals.

Data Analysis

Data analysis followed a thematic approach as recommended by Braun and Clarke (2006). First, the interview transcripts and other qualitative data were systematically coded to identify key themes and patterns. The steps included familiarizing with the data by reviewing transcripts, generating initial codes based on key phrases and concepts, and grouping codes into broader themes relevant to AI-driven green marketing. These themes were then refined to ensure they aligned with the research questions. The final themes provided a coherent narrative that shed light on the use of AI in promoting green marketing and sustainability in Harare. Representative quotes and examples were selected to illustrate the core findings, ensuring the study provides actionable insights for businesses and policymakers in Zimbabwe. Focusing on a localized context, this study aims to offer a detailed exploration of how businesses in Harare can leverage AI to drive green marketing and contribute to the broader goals of sustainability and the circular economy. The emphasis on qualitative methods allows for a deep understanding of the practical applications and challenges associated with AI adoption in Zimbabwe's unique market conditions.

Results

The qualitative data collected from interviews with business leaders, marketing professionals, and industry experts provide rich insights into the application of AI-driven green marketing strategies in African markets. The thematic analysis of the interview data revealed three major themes: (1) Optimization of Resource Efficiency, (2) Targeted Consumer Engagement, and (3) Barriers to AI Adoption. Each of these themes offers a deeper understanding of how AI technologies are reshaping marketing strategies to promote sustainability within the context of African economies (Table 11.1).

Optimization of Resource Efficiency

The interviews revealed a recurring theme of optimizing resource efficiency through the integration of AI into green marketing strategies. Participants frequently emphasized the transformative role of AI technologies, such as predictive analytics and machine learning, in reducing waste and enhancing resource management. Across various industries, including agriculture and energy, interviewees noted that AI-driven systems enabled more precise forecasting of resource needs, which led to a significant reduction in resource wastage while advancing sustainability goals. One participant from a Nigerian agribusiness explained, "*AI has helped us reduce water and fertiliser usage by accurately predicting the needs of our crops based on weather patterns and soil conditions. This has significantly lowered our operational costs while promoting sustainability.*" Similarly, a representative from the energy sector noted that AI allowed their company to "optimise energy consumption by predicting peak demand periods, ensuring that resources are used efficiently and sustainably." These sentiments reinforce the study's theoretical foundations, particularly sustainability theory and the circular economy, where efficient resource management is a core principle in achieving broader environmental objectives.

Table 11.1 Thematic analysis of AI-driven green marketing strategies in African markets

Main themes	*Sub-themes*	*Description*
1. Optimization of resource efficiency	a. AI-driven analytics	Utilization of AI tools to analyze data for improved resource management and waste reduction.
	b. Predictive modeling	Employing predictive analytics to forecast demand and optimize inventory levels.
	c. Supply chain management	Enhancements in tracking and managing supply chains for sustainable practices.
	d. Energy consumption reduction	Strategies to minimize energy use through AI technologies in production and logistics.
2. Targeted consumer engagement	a. Personalized marketing	Tailoring marketing strategies based on consumer behavior and preferences for eco-friendly products.
	b. Consumer segmentation	Leveraging AI to identify and segment consumers interested in sustainable practices.
	c. Social media insights	Using AI to analyze social media data for consumer sentiment regarding sustainability.
	d. Building consumer trust	Strategies to enhance trust through transparency in sourcing and sustainability efforts.
3. Barriers to AI adoption	a. Technological infrastructure disparities	Challenges related to varying levels of technology access and infrastructure in African markets.
	b. Consumer awareness	The need for increased consumer education on the benefits of AI and sustainable practices.
	c. Regulatory and policy challenges	Existing policies that may hinder the adoption of AI in green marketing strategies.
	d. Financial constraints	Limitations in funding and investment for AI technologies among businesses in emerging markets.

Targeted Consumer Engagement

The second prominent theme that emerged from the interviews was the use of AI to enhance targeted consumer engagement in green marketing campaigns. Participants consistently reported that AI technologies such as Natural Language Processing (NLP) and machine learning had fundamentally transformed how they segmented their audience and personalized marketing messages. Interviewees described how these AI-driven tools allowed them to identify and engage eco-conscious consumers with more precision. For instance, an interviewee from a South African retail company highlighted the benefits of AI-driven customer segmentation, explaining, "AI helps us understand our customers' environmental concerns and preferences, which enables us to create marketing campaigns that resonate with their values. This has significantly increased customer engagement and loyalty towards our sustainable product lines." Similar experiences were shared by participants across different sectors, indicating that businesses adopting AI for targeted engagement experienced a marked increase in consumer interaction with eco-friendly products. These findings strongly reinforce the theoretical frameworks of marketing theory, particularly the value-belief-norm (VBN) theory, which posits that aligning consumer values with pro-environmental behaviors leads to greater engagement and support for sustainable practices.

Barriers to AI Adoption

Despite the optimism surrounding AI's potential to drive green marketing, a recurring theme in the interviews was the significant barriers to AI adoption, particularly in the resource-constrained environments common across many African markets. Participants frequently highlighted challenges such as the high costs of implementing AI technologies, limited access to quality data, and the lack of technical expertise. For instance, an interviewee from a Kenyan startup remarked, "AI has immense potential, but the initial investment is a significant hurdle for smaller companies like ours. Additionally, we struggle with obtaining the quality data needed to train our AI models effectively." Several participants echoed these concerns, further pointing to the uneven technological infrastructure across Africa, which impedes the widespread adoption of AI-driven green marketing strategies. These obstacles underscore the necessity for supportive policies and investment in infrastructure to ensure the scalability of AI technologies in promoting green marketing initiatives across the continent.

Discussion

The findings of this study offer valuable insights into the practical applications of AI-driven green marketing strategies within African markets. Analyzing the themes that emerged from the interviews, optimization of resource efficiency, targeted consumer engagement, and barriers to AI adoption, the study critically compares results with the existing literature, highlighting both alignment and gaps.

Optimization of Resource Efficiency

The study reveals that businesses in African markets are increasingly leveraging AI technologies to optimize resource efficiency, particularly in sectors such as agriculture and energy. These results align with existing studies that emphasize AI's potential to enhance sustainability efforts through improved resource management. Patel and Singh (2023) highlight AI's ability to streamline resource allocation and minimize waste, thus leading to the more efficient use of water, energy, and raw materials. Similarly, Lacy and Rutqvist (2021) discuss how AI can optimize supply chain processes, reducing environmental impacts and promoting circular economy principles. Huang et al. (2021) also underscore AI's role in improving resource efficiency across various industries by predicting demand and minimizing wastage. While much of the literature focuses on resource efficiency in developed markets, our findings offer unique insights by contextualizing these benefits within African markets. AI's ability to predict resource needs such as water and fertilizer usage in agriculture demonstrates its value in regions facing resource scarcity. This corroborates the arguments of Nwankwo and Mbachu (2022), who suggest that AI can mitigate environmental and resource challenges that are particularly acute in African economies. Huang et al. (2021) found that AI-driven predictive analytics in Ghanaian agribusinesses significantly improved resource management, enhancing sustainability practices. Our study extends these insights, demonstrating that AI-driven resource optimization is not a mere theoretical concept but a tangible outcome observable in real-world African business practices. These applications underscore AI's potential to drive both sustainability and economic growth, even in regions facing significant environmental constraints (Abiodun et al., 2021).

Targeted Consumer Engagement

The findings on AI's role in enhancing targeted consumer engagement are consistent with existing literature, which consistently highlights AI's capacity to personalize marketing efforts and influence consumer behavior. Emon and Khan (2024) demonstrate that AI technologies enable businesses to develop more targeted and personalized campaigns, which increase customer engagement and loyalty. The use of AI-driven customer segmentation observed in this study supports these findings, showing that businesses are effectively tailoring marketing messages to eco-conscious consumers, thereby driving higher engagement with sustainable products. This aligns with Gupta and Sharma (2023), who found that AI-driven personalization significantly boosts consumer interaction, particularly in sectors promoting green products. The results corroborate the theoretical framework of the value-belief-norm (VBN) theory, as discussed by Stern (2000). According to this theory, consumers are more likely to engage in sustainable behaviors when marketing strategies align with their environmental values and beliefs. Our study supports this, showing that AI's ability to personalize marketing based on consumer preferences and values leads to greater interaction with eco-friendly products. Zhang et al. (2023) similarly demonstrated that AI helps businesses identify new market opportunities for eco-friendly products, thus driving both business growth and sustainability efforts. This is further echoed by Hinson et al. (2021), who found that AI-driven targeted marketing plays a pivotal role in fostering customer loyalty towards sustainable brands. This study also highlights a contextual gap in the literature. Most existing research centers on developed markets, where consumer behavior regarding sustainability is relatively more mature. The insights from this study offer novel perspectives on how AI can facilitate a shift towards sustainable consumption in African markets, where consumer awareness of environmental issues is still emerging. As shown in the findings, AI-driven strategies helped businesses in Africa engage consumers by tailoring messages to their growing environmental consciousness, an area often overlooked in previous research (Adeola & Ezenwafor, 2021). This suggests that AI not only influences consumer engagement in developed economies but also plays a critical role in nurturing pro-environmental behaviors in less mature markets, reinforcing its transformative potential in promoting sustainability across diverse contexts (Emon & Khan, 2024).

Barriers to AI Adoption

A critical insight from the study is the identification of significant barriers to AI adoption in African markets. While the literature widely recognizes AI's transformative potential in enhancing green marketing (Emon & Khan, 2024), our findings shed light on the substantial challenges faced by businesses in Africa. These challenges include high implementation costs, limited access to quality data, and a shortage of technical expertise, all of which hinder the widespread adoption of AI technologies. This observation echoes the concerns raised by Adeola and Ezenwafor (2021), who noted that resource-constrained environments in Africa often limit businesses' ability to fully capitalize on AI's potential. Hinson et al. (2021) and Mukonza et al. (2021) highlighted the technological and infrastructural limitations that prevent African businesses from implementing advanced AI-driven solutions. The study confirms these limitations; it also extends the conversation by emphasizing the uneven technological landscape across different African regions. Some areas possess the infrastructure to support AI, while others lag behind, making it difficult to scale AI-driven green marketing efforts uniformly across the continent. This observation aligns with the work of Chimuka (2022), who discussed the disparity in AI adoption due to regional infrastructural imbalances in Africa. The barriers identified in our study contrast with research conducted in more technologically advanced markets, where the primary focus is on maximizing AI's potential rather than overcoming infrastructural or financial hurdles (Chen & Chang, 2021). These findings contribute to a deeper understanding of how context-specific factors in African markets shape the adoption and effectiveness of AI-driven green marketing strategies. This underscores the critical need for tailored policies and investments that address the unique challenges faced by businesses in Africa, ensuring that AI's potential can be fully realized in promoting sustainable practices across the continent.

Comparing the Results with Sustainability Theory

The integration of AI into green marketing strategies closely aligns with the principles of sustainability theory, particularly the triple bottom line (Elkington, 1997) and the circular economy (Ellen MacArthur Foundation, 2013). The study demonstrates that AI technologies enable businesses in Africa to balance economic, social, and environmental objectives, thereby supporting the argument that sustainability requires a holistic approach extending beyond financial success. These findings are consistent with Lacy and Rutqvist (2021), who emphasize AI's role in driving circular economy initiatives by optimizing resource flows and reducing waste, thus contributing to more sustainable business models. While sustainability theory often assumes that businesses have the resources and capabilities to adopt these models, our findings reveal a more complex reality in African markets. Many businesses struggle with the financial and technical challenges associated with AI adoption, making it difficult to fully realize the potential of the circular economy. This concern mirrors the observations of Adeola and Ezenwafor (2021), who noted that the resource limitations in emerging economies often hinder the implementation of advanced sustainability practices. Studies by Patel and Singh (2023) highlight that the high costs of AI technologies and the lack of technical expertise pose significant barriers to achieving sustainable outcomes in these regions. The study identifies a gap in the existing literature, which frequently overlooks the specific barriers faced by businesses in emerging economies. The assumption that all businesses can easily adopt sustainability models like the circular economy fails to consider the unique challenges present in African markets. Pinpointing these challenges, our research suggests that tailored policies, financial support, and capacity-building initiatives are essential for enabling African businesses to effectively adopt AI-driven sustainability practices. This expands on the work of Geissdoerfer et al. (2020), who argue for more inclusive approaches to sustainability that account for the varying capacities of businesses across different regions and economies.

Implications for Theory

The study expands the theoretical framework of green marketing through the integration of AI principles, sustainability theory, and marketing theory. It offers a nuanced perspective on how AI-driven strategies can enhance green marketing efforts and foster a circular economy. Contextualizing these theories within the specific challenges of African markets highlights the importance of regional and socio-economic factors when applying AI for sustainability. This interdisciplinary approach enriches the existing literature and opens new avenues for research at the intersection of AI, green marketing, and sustainable development.

Implications for Policy

The findings highlight the need for policymakers to support the adoption of AI technologies in green marketing as a pathway to achieving sustainable development goals. Policies should promote investment in AI infrastructure and offer incentives for businesses to integrate AI-driven sustainability practices. Creating regulatory frameworks that ensure ethical AI use, protect consumer data, and promote transparency in AI applications is equally crucial. Facilitating AI adoption in green marketing will help policymakers drive innovation, foster economic growth, and address environmental challenges.

Implications for Practice

The study offers practical insights into using AI to enhance green marketing strategies. Adopting AI technologies like machine learning, predictive analytics, and natural language processing enables companies to develop more targeted and personalized marketing campaigns that resonate with eco-conscious consumers. AI's role in optimizing resource management, reducing waste, and improving operational efficiency provides businesses with a competitive advantage, increased consumer engagement, and stronger brand loyalty. The research also delivers actionable recommendations for startups and established firms in African markets, encouraging innovation and helping them achieve sustainability goals through AI-driven green marketing.

Limitations and Areas for Further Research

While this study offers valuable insights into AI-driven green marketing strategies in African markets, several limitations must be acknowledged. The research relies on qualitative data from interviews, which, though providing in-depth perspectives, are limited by the subjectivity of participants and the small sample size, potentially reducing the findings' applicability across all industries in Africa. Expanding future research with larger, more diverse samples and incorporating both qualitative and quantitative methods could yield more generalizable results. The study's geographic focus also does not cover the full diversity of African markets, meaning the findings may not apply uniformly across the continent. Technological and infrastructural constraints in many African markets hinder businesses from fully accessing and implementing advanced AI technologies. Future research should explore strategies to overcome these barriers and examine how businesses in resource-constrained environments adapt AI to their needs. Lastly, given the rapid pace of AI advancements, continuous updates and further studies are needed to track how these technologies shape green marketing and sustainability efforts over time.

Future research should continue to investigate how AI-driven green marketing strategies can be adapted to the unique economic and social contexts of different African regions and industries. Exploring solutions to the challenges of data accessibility, infrastructure limitations, and cost barriers will be essential in ensuring the widespread adoption of AI technologies. There is a need to examine how AI can further enhance the circular economy by integrating more advanced technologies such as the Internet of Things (IoT) and blockchain for better transparency and resource tracking. Longitudinal studies could also provide insights into the long-term impact of AI-driven green marketing on both business performance and sustainability outcomes. Addressing these areas, future research can help ensure that AI-driven strategies not only promote sustainability but also contribute to equitable economic growth across Africa.

Conclusion

This study has explored the transformative potential of AI-driven green marketing strategies within the context of African markets, focusing on their role in promoting a circular economy. Through thematic analysis, three key areas emerged: optimization of resource efficiency, targeted consumer engagement, and barriers to AI adoption. The findings demonstrate that AI technologies, such as predictive analytics and machine learning, can significantly enhance sustainability efforts by reducing waste, improving resource management, and fostering more personalized marketing that resonates with eco-conscious consumers. These outcomes align with existing sustainability and marketing theories, reinforcing the notion that AI can be a powerful tool for driving sustainable business practices. Despite the clear benefits, several challenges remain that hinder the broader adoption of AI-driven green marketing strategies across African markets. High implementation costs, limited access to quality data, and uneven technological infrastructure continue to pose significant barriers, especially for smaller businesses. These challenges highlight the need for targeted policy interventions and investments to bridge the gap between AI potential and its practical application. Governments, businesses, and other stakeholders must collaborate to create supportive frameworks that address these obstacles, allowing African businesses to fully capitalize on the potential of AI-driven sustainability initiatives.

References

Abiodun, R., Eze, C., & Omisanya, T. (2021). AI applications in African markets: Sustainability and resource optimization. *Journal of African Business and Sustainability, 23*(1), 78–94.

Achebe, C. (2020). Sustainable innovations in African agriculture: AI-driven approaches. *International Journal of Sustainable Development, 18*(2), 150–165.

Adeola, A., & Ezenwafor, P. (2021). Challenges and opportunities in AI-driven green marketing in Africa. *Journal of Business and Technology in Africa, 19*(3), 102–120.

Adetunji, A., & Oke, P. (2020). The potential of AI for sustainability in emerging markets: A review of applications in Africa. *Sustainability Review, 14*(2), 134–153.

Agrawal, R., Wankhede, V. A., Kumar, A., Upadhyay, A., & Garza-Reyes, J. A. (2022). Nexus of circular economy and sustainable business performance in the era of digitalization. *International Journal of Productivity and Performance Management, 71*(3), 748–774.

Agyeman, J., Boateng, D., & Oppong, R. (2021). Consumer education in green marketing: Implications for sustainability. *Journal of Green Marketing, 16*(3), 85–101.

Akira AI. (2023). AI agents for sustainability and waste reduction. Available at: https://www.akira.ai/blog/ai-agents-for-sustainability-and-waste-reduction (Accessed: 18 January 2025).

Bandura, A. (1986). Social foundations of thought and action. Englewood Cliffs, NJ (23–28), 2.

Belanche, D., Casaló, L. V., Flavián, M., & Ibáñez-Sánchez, S. (2021). Understanding influencer marketing: The role of congruence between influencers, products and consumers. *Journal of Business Research, 132*, 186–195.

Bhardwaj, A. K., Garg, A., Ram, S., Gajpal, Y., & Zheng, C. (2020). Research trends in green product for environment: A bibliometric perspective. *International Journal of Environmental Research and Public Health, 17*(22), 8469.

Bocken, N. M., Weissbrod, I., & Antikainen, M. (2021). Business model experimentation for the circular economy: Definition and approaches. *Circular Economy and Sustainability, 1*(1), 49–81.

Braun, V., & Clarke, V. (2006). Using thematic analysis in psychology. *Qualitative Research in Psychology, 3*(2), 77–101.

Chen, Y., & Chang, R. (2021). Drivers and barriers to green marketing adoption: A comparative study. *Journal of Marketing and Sustainability, 24*(4), 60–77.

Chimuka, G. (2022). The role of the media and the international community in recent conservation issues in Southern Africa: The case of Cecil the lion. In *Protected areas and tourism in Southern Africa* (pp. 146–160). Routledge.

Choi, S., & Kim, I. (2021). Sustainability of nature walking trails: Predicting walking tourists' engagement in proenvironmental behaviors. *Asia Pacific Journal of Tourism Research, 26*(7), 748–767.

Christensen, C. M., McDonald, R., Altman, E. J., & Palmer, J. E. (2018). Disruptive innovation: An intellectual history and directions for future research. *Journal of Management Studies, 55*(7), 1043–1078.

Dwivedi, Y. K., Hughes, D. L., Baabdullah, A. M., & Ribeiro, R. (2021). AI and green marketing: Emerging trends and applications. *International Journal of Information Management, 57*, 102261.

Elkington, J. (1997). *Cannibals with forks: The triple bottom line of 21st-century business.* Capstone Publishing.

Ellen MacArthur Foundation. (2013). *Towards the circular economy: Economic and business rationale for an accelerated transition.* Ellen MacArthur Foundation.

Ellen MacArthur Foundation. (2019). Artificial Intelligence and the Circular Economy Ellen MacArthur Foundation, Artificial intelligence and the circular economy - AI as a tool to accelerate the transition. http://www.ellenmacarthurfoundation.org/publications

Emon, M. M. H., & Khan, T. (2024). A systematic literature review on sustainability integration and marketing intelligence in the era of artificial intelligence. *Review of Business and Economics Studies, 12*(4), 6–28.

Evans, P., & Cooper, L. (2022). AI-driven circular economy: Reducing waste through technology. *Circular Economy Insights, 9*(3), 115–132.

Geissdoerfer, M., Savaget, P., Bocken, N. M. P., & Hultink, E. J. (2020). The circular economy: A new sustainability paradigm? *Journal of Cleaner Production, 143*, 757–768.

Geissdoerfer, M., Vladimirova, D., & Evans, S. (2018). Sustainable business model innovation: A review. *Journal of cleaner production, 198*, 401–416.

Gupta, A., & Sharma, P. (2023). Personalised green marketing strategies using AI. *Green Business Journal, 22*(1), 45–59.

Hinson, R., Boateng, D., & Yeboah, A. (2021). Sustainability marketing in African markets: Bridging the technology gap. *African Journal of Business Ethics, 13*(2), 56–75.

Huang, X., Lee, Y., & Wang, H. (2021). AI-driven sustainability practices: Optimizing resources through technology. *Technological Forecasting and Social Change, 173*, 120912.

Jiang, Q., Bhattarai, N., Pahlow, M., & Xu, Z. (2022). Environmental sustainability and footprints of global aquaculture. *Resources, Conservation and Recycling, 180*, 106183.

Kaplan, N. E., Baker, K. S., & Karasti, H. (2021). Long live the data! Embedded data management at a long-term ecological research site. *Ecosphere, 12*(5), e03493.

Kim, H., & Park, M. (2023). Virtual influencers' attractiveness effect on purchase intention: A moderated mediation model of the Product–Endorser fit with the brand. *Computers in Human Behavior, 143*, 107703.

Kumar, S., & Prakash, R. (2020). Green marketing principles and sustainability practices. *Journal of Sustainable Marketing, 19*(1), 23–41.

Kumar, S., Lim, W. M., Sivarajah, U., & Kaur, J. (2023). Artificial intelligence and blockchain integration in business: Trends from a bibliometric-content analysis. *Information Systems Frontiers, 25*(2), 871–896.

Lacy, P., & Rutqvist, J. (2021). AI in the circular economy: Enhancing supply chains and reducing waste. *Sustainability Science and Business, 29*(3), 90–105.

Liu, J., Zhao, D., Mao, G., Cui, W., Chen, H., & Yang, H. (2020). Environmental sustainability of water footprint in mainland China. *Geography and Sustainability, 1*(1), 8–17.

Luo, X., & Wilson, D. (2021). AI's influence on consumer knowledge and behavior in green marketing. *Journal of Marketing Technology, 28*(1), 64–79.

Mensah, H. K., Agyapong, A., & Osei, B. A. (2021). Effect of corporate social responsibility on eco-citizenship behaviour in luxury hotels: Eco-lifestyle as a moderator. *Journal of Global Responsibility, 12*(2), 189–209.

Mishra, M., Hota, S. K., Ghosh, S. K., & Sarkar, B. (2020). Controlling waste and carbon emission for a sustainable closed-loop supply chain management under a cap-and-trade strategy. *Mathematics, 8*(4), 466.

Mishra, V., & Sung, C. (2022). Machine learning's impact on green marketing and resource optimization. *Environmental Economics and AI, 34*(4), 112–124.

Moyo, B., & Ndlovu, N. (2023). Examining the challenges and survival strategies of the informal sector during COVID-19 induced lockdowns and restrictions in Bulawayo.

Mukonza, C., Hinson, R. E., Adeola, O., Adisa, I., Mogaji, E., & Kirgiz, A. C. (2021). Green marketing: An introduction. *Green Marketing in Emerging Markets: Strategic and Operational Perspectives*, 3–14.

Murray, A., Skene, K., & Haynes, K. (2017). The circular economy: An interdisciplinary exploration of the concept and application in a global context. *Journal of Business Ethics, 140*, 369–380.

Mwangi, J. (2024). Analyzing the role of artificial intelligence and machine learning in optimizing supply chain processes in Kenya. *International Journal of Supply Chain Management, 9*(1), 39–50.

Mwenda, E., Muange, E. N., Ngigi, M. W., & Kosgei, A. (2023). Impact of ICT-based pest information services on tomato pest management practices in the Central Highlands of Kenya. *Sustainable Technology and Entrepreneurship, 2*(2), 100036.

Nwankwo, P., & Mbachu, E. (2022). AI-driven green marketing strategies in Nigeria: Opportunities and challenges. *Journal of Emerging Markets, 17*(2), 84–101.

Nzomo, J. K., Kinyua, G. M., & Mwasiaji, E. T. (2023). Green innovation strategy as antecedent of performance sustainability among ISO 14001 certified manufacturing firms in Kenya: Does competitive advantage play a mediating role. *International Journal of Education and Research, 11*(8), 93–108.

Obaidi, M., Nagel, L., Specht, A., & Klünder, J. (2022). Sentiment analysis tools in software engineering: A systematic mapping study. *Information and Software Technology, 151*, 107018.

Okolo-obasi, N. E., & Uduji, J. I. (2023). Does agri-business/small and medium enterprise investment scheme (AGSMEIS) impact on youth entrepreneurship development in sub-Saharan Africa? Evidence from Nigeria. *Journal of Economic and Administrative Sciences, 39*(3), 571–595.

Okoro, C., & Nwachukwu, A. (2022). Oil formation volume factor prediction using artificial neural network: A case study of niger delta crudes. *Journal of Earth Energy Engineering, 11*(3), 125–136.

Patel, S., & Singh, A. (2023). AI's role in sustainable marketing: Economic and environmental impacts. *International Journal of Sustainability Studies, 21*(2), 45–63.

Rogers, E. (2003). *Diffusion of innovations*. Free Press.

Stern, P. C. (2000). New environmental theories: Toward a coherent theory of environmentally significant behavior. *Journal of Social Issues, 56*(3), 407–424.

Sun, S., Sang, W., & Axmacher, J. C. (2020). China's national nature reserve network shows great imbalances in conserving the country's mega-diverse vegetation. *Science of The Total Environment, 717*, 137159.

Tseng, M. L., Chang, C. H., Lin, C. W. R., Wu, K. J., Chen, Q., Xia, L., & Xue, B. (2020). Future trends and guidance for the triple bottom line and sustainability: A data driven bibliometric analysis. *Environmental Science and Pollution Research, 27*, 33543–33567.

Wang, Z., Huang, F., Liu, J., Shuai, J., & Shuai, C. (2020). Does solar PV bring a sustainable future to the poor?—An empirical study of anti-poverty policy effects on environmental sustainability in rural China. *Energy Policy, 145*, 111723.

Witjes, S., Lozano, R. (2016). Towards a more Circular Economy: Proposing a framework linking sustainable public procurement and sustainable business models. *Resources, Conservation and Recycling, 112*, 37–44.

Yadav, A., & Gupta, M. (2021). Leveraging AI for sustainability in emerging economies. *Journal of Sustainable AI Development, 10*(2), 140–157.

Zhang, X., Wang, L., Helwig, J., Luo, Y., Fu, C., Xie, Y., ... & Ji, S. (2023). Artificial intelligence for science in quantum, atomistic, and continuum systems. arXiv preprint arXiv:2307.08423.

CHAPTER 12

Disseminating Sustainable End-of-life Product Disposal Information Through Social Media Nudging

Paul Mukucha, Tatenda Katsande, Thulani Dube, Divaries Cosmas Jaravaza, and Sharon Chisango

Introduction

The invention of single-use baby diapers is one of the greatest innovations that alleviated the baby caregiving burden (Randles, 2021). Most African cultures impose such a burden solely on women. Traditionally, women are

P. Mukucha (✉) • T. Dube • D. C. Jaravaza (✉)
Department of Marketing, Bindura University of Science Education, Bindura, Zimbabwe
e-mail: pmukucha@buse.ac.zw; tdube@buse.ac.zw; djaravaza@buse.ac.zw

T. Katsande
Business Management Department, Lupane State University, Bulawayo, Zimbabwe
e-mail: tkatsande@lsu.ac.zw

A. Muposhi, B. Nyagadza (eds.), *Sustainable Green Marketing Strategies for a Circular Economy*, Palgrave Studies of Marketing in Emerging Economies,
https://doi.org/10.1007/978-3-031-81627-7_12

socialized to conduct baby caregiving duties that involve changing soiled diapers among other demanding domestic chores (Waterhouse et al., 2017). These duties were quite understandable during the era when most women were excluded from both formal and informal employment sectors (Nzima & Maharaj, 2023). However, as most women are increasingly formally employed, adjustments had to be put in place with regard to their baby-caring practices, particularly involving the use and disposal of diapers (Luthuli et al., 2020). The emergence of single-use baby diapers then came as a relief for the burdened mothers (Płotka-Wasylka et al., 2022). Henceforth, currently in Zimbabwe, the majority of the baby diapers in the market are designed for single use (Mangizvo et al., 2024), and babies use an average of five diapers per day (Ntekpe et al., 2020). This suggests higher volumes and frequencies of disposals per day.

Recently, the disposal of single-use products in general has been classified as a growing environmental threat (Płotka-Wasylka et al., 2022; Mukucha et al., 2023a, 2023b). This has particularly been more pronounced in the disposal of single-use baby diapers. The threat came as a result of most baby caregivers not adhering to the widely acclaimed standards for dealing with single-use diapers. Single-use diapers are supposed to be separated from their contents before they are thrown away (Płotka-Wasylka et al., 2022). More specifically, the contents of single-use baby diapers are supposed to be thrown into a toilet and then the diapers are washed and sealed before being thrown away into a waste material bin (Ntekpe et al., 2020). The residual waste material, if improperly handled, may lead to water pollution that is associated with the spread of diseases such as *escherichia coli* and cholera (Ntekpe et al., 2020; Narayan et al., 2022). After sealing, the soiled diaper is supposed to be placed in a plastic paper that should be sealed as well in order to eliminate any odour that may cause air pollution (Hussain et al., 2017).

There is a mushrooming of dumpsites littered with used baby diapers that are simply discarded without practicing the separation at source (White et al., 2023; Muringaniza et al., 2023; Schenck et al., 2023). Separation at source entails separating fecal matter from the used diapers before discarding the used diapers into a trash bin (Kuhl et al., 2021). It is most likely that most mothers may not be privy to the acceptable baby

S. Chisango
Department of Human Capital Management, Bindura University of Science Education, Bindura, Zimbabwe
e-mail: schisango@buse.ac.zw

diapers disposal practices (Makoś-Chełstowska et al., 2021; Sachidhanandham & Priyanka, 2020). This has raised issues related to compliance with the United Nations Sustainable Development Goals (UNSDG). The promotion of safe disposal of single-use baby diapers is enshrined in several SDGs, such as SDG6, 11, and 12. SDG6 relates to clean water and sanitation. Irresponsible disposal of used baby diapers leads to drinking water contamination that results in waterborne diseases. SDG11 advocates for sustainable cities and communities, which means promoting a healthy environment by putting in place proper waste collection and treatment systems. This involves avoiding indiscriminate disposal of pollutants such as used baby diapers. Finally, SDG12 promotes responsible production and consumption. With regard to used baby diapers, disposal processes should ensure that there is proper waste separation at source to enable safe disposal of pollutants and recycling processes.

Various health professionals such as midwives and nurses regularly educate women on responsible disposal of used baby diapers through various mainstream media platforms such as radios and television (Friedli, 2017; Hudnut-Beumler et al., 2016). Such information is more likely to reach the targeted audience if it is conveyed through social media since most people spend most of their time on social media for entertainment and vital information. In this regard, social media skits have become the prime method of communication amongst most societies on the basis that they easily get shared at a very affordable cost and the audience has control over the timing of message reception (Dangaiso et al., 2024). Previous research has proved the efficacy of messages promoted through the social media skits (e.g. Jaravaza et al., 2024). However, a search in the extant literature did not reveal any application of social media skits to convey messages related to consumption and disposal of personal care products such as single-use baby diapers. Therefore, the primary objective of this chapter is to determine the efficacy of environmental consciousness messages conveyed through social media skits in eliciting waste materials separation at source awareness, attitudes, and behavior.

The rest of this chapter is structured as follows: Literature review leveraging on the nudge framework as an overarching theory for linking social media skits to various waste material separation outcomes such as awareness, attitudes, and behaviur. Literature review shall be followed by a methodology section that outlines how the data collection and analysis processes were conducted. This shall be followed by the results presentation and discussion sections.

Literature Review

Promoting Pro-environmental Behavior Using Social Media Skits

Social media is defined as an interaction between individuals through virtual platforms such as WhatsApp, TikTok, Twitter, and Facebook (Aichner et al., 2021; Evans et al., 2021; Carr & Hayes, 2015). Social media is characterized by virtual communities and networks that enable the sharing of multimedia content (Tuten, 2023; Evans et al., 2021). The use of social media has led to the serendipitous emergence of social media skits. A social media skit is defined as a creative and entertaining short video for promoting specific ideas (Ojomo & Sodeinde, 2021). Social media skits draw the attention of viewers through employing humor, satire, and parodies (Afolabi & Falusi, 2023). They encourage interaction through allowing for feedback in the form of likes, shares, and comments (Faltesek et al., 2023; Ling et al., 2022).

Social media skits have developed into innovative and effective tools for disseminating environmental consciousness promotional messages (e.g. Da Mota & Pickering, 2020). This arises from their ability to reach diverse and wide audiences (Divya et al., 2023), as well as facilitating social engagements (Atherton, 2023). Previous research has shown that social media skits interventions increase awareness (Ortiz et al., 2019). They also influence attitude formation (Langaro et al., 2018) and behavioral change (Schivinski et al., 2019). However, while social media skits have shown to have some efficacy in influencing various psychological constructs, to the best knowledge of these researchers, there are no specific studies that indicated how social media skits influence environmentally related levels of awareness, attitudes, and behaviors. There are some challenges associated with social media skits usage (Martí et al., 2019; Li et al., 2019). These challenges relate to assessing evaluation mechanisms. The effectiveness of social media skit interventions as a communication strategy is still hindered by less developed social media communication strategies, unsuitable measurement research designs, message misinformation, and inabilities to measure long-term impact (Huang et al., 2022; Kizgin et al., 2020).

Waste Material Separation at Source

Waste material separation at source is a sustainable practice of conducting solid waste management (Khanal et al., 2023). This involves separating various types of waste material into their distinct categories for the purposes of easy transportation, disposal, and recycling (Le Dinh et al., 2021). Waste material separation for the purposes of effective disposal is particularly important for used baby diapers (Purwati et al., 2021). The generally acceptable practice of waste separation for used baby diapers involves separating the fecal matter and disposing it in an ablation facility (Reese et al., 2015) and then securely wrapping the used diaper in a suitable disposal bag in order to prevent any leakages and the usual odours (Ntekpe et al., 2020; Mathe, 2018). The sealed diaper disposable bag is then disposed in line with the prescribed local regulations (Wambui et al., 2015).

The waste material separation at source as a disposal method for used baby diapers is the primary activity in preparation for reusing reusable diapers and recycling single-use baby diapers (Płotka-Wasylka et al., 2022). Recently, there has been some inventions for recycling single-use baby diapers (Itsubo et al., 2020; Khoo et al., 2019; Mendoza et al., 2019). The recycling of single-use baby diapers has a positive environmental impact of reducing all kinds of pollution (Sachidhanandham & Priyanka, 2020).

Hypotheses Development

The link between social media skit messages and waste material separation at source can be accounted by the nudging concept as an overarching theory. The nudge theory was conceptualized by Thaler and Sunstein (2021), and it posits that choices that influence individuals to make certain decisions are a function of subtle changes introduced around options presented to them. The theory suggests that individuals should be nudged towards opting for better choices. When applied to environmental preservation initiatives, the nudge theory can help to identify interventions that can lead to orthodox changes (Barker et al., 2021; John et al., 2019). One such intervention that has gained widespread acceptance and greater popularity is nudging through social media skits (Mirbabaie et al., 2021) in what is referred to as digital nudging (Monge Roffarello & De Russis, 2023).

Social media skits can convey environmentally related messages that can nudge the recipients to adopting sustainable choices, follow social norms, and attain feedback loops for positive reinforcement. Furthermore, social skits can also convey personalization of message content and the provision of incentives to encourage the promoted attitudes and behaviors (Ren &

Liu, 2024; Kroll & Stieglitz, 2021; Bammert et al., 2020). This is normally achieved through content that is laced with some overtones of entertainment (Ojomo & Sodeinde, 2021).

Waste Material Separation Awareness

Waste material separation at source awareness campaigns are important for households since households are the sources of most waste material (Agbefe et al., 2019). The campaigns can be executed through social media skits. The influence of social media skits on enhancing waste separation awareness can be accounted by the social amplification theory (SAT) conceptualized by Kasperson et al. (1988). This theory suggests that there is some amplification of the information shared through social media through several behavioral user-engagement mechanisms that include, but are not limited to, shares, likes, retweets, and comments (Kasperson et al., 2022; Wirz et al., 2018). Such mechanisms of information shared through social media lead to increases in the levels of visibility and reach (Hasell, 2021). The theory also posits that social media information flow is influenced by factors ranging from personal experiences, social networks, media coverage, and cultural norms (Zhang & Cozma, 2022).

The basic elements of SAT include amplification stations for increasing message reach, reception processes for creating interest in the messages, cognitive responses, and secondary reactions that amplify the original message. More so, the theory also includes social networks that moderate how information gets viral within a community (Coombs & Holladay, 2021; Chong & Choy, 2018; Fellenor et al., 2018). In the context of raising awareness through social media, the SAT describes how the shared information gains momentum as it spreads in various communities, leading to higher levels of visibility (Zhao & Wu, 2021; Ng et al., 2018). Previous research has indicated that social media skits have some efficacy in attaining some form of awareness (Iyanuoluwa & Oyinlade, 2024; Sheak & Abdulrazak, 2023). Therefore, drawing from the nudge theory and SAT, it is plausible to suggest that

H1 Social media skits have some efficacy in influencing waste material separation awareness

Waste Material Separation Attitudes

Social media skits are believed to have an influence on individuals' attitudes (Omoyajowo et al., 2024; Afolabi & Falusi, 2023; Okoye & Okenyi, 2023). The shaping of attitudes through social media skits can be accounted by the looking glass self-theory. The looking glass self-theory was proposed by Cooley (1902) to describe how individuals develop a sense of self on the basis of how they believe to be perceived by others (Ruiz-Junco & Brossard, 2019). More specifically, this theory suggests that individuals gauge their own worth and attitudes through social interaction as a mirror (Ruiz-Junco & Brossard, 2019; Zhao, 2018). In the case of social media skit usage, individuals tend to get various messages by interacting with others, who in most cases are their peers or people of the same social structure.

While social media skits have some influence on shaping individuals' attitudes, individuals may maintain their original attitudes through unintended strategies such as selective exposure, counter arguing, and attitude bolstering (Chen et al., 2022; Vafeiadis et al., 2020; Song & Boomgaarden, 2017). This may involve seeking information aligned to their existing beliefs and disregarding misaligned information posted through social media skits (Braddock & Morrison, 2021; Post, 2019). Nevertheless, on a balance of probabilities, emanating from the nudge theory and the looking glass self-theory, it is anticipated that

H2 Social media skits have some efficacy in influencing waste material separation attitudes

Waste Material Separation Behavior

Social media skits are likely to have an influence on waste separation practices. This influence can be explained better through the lens of the social cognitive theory. The social cognitive theory was conceptualized by Bandura (2001) to explain the role of observational learning as well as imitation in shaping individuals' behaviors. The theory suggests that individuals' behavior is influenced by the interaction between personal and environmental factors (Schunk & DiBenedetto, 2020; Beauchamp et al., 2019). The theory further recognizes the role of reinforcements in shaping behavior (Bandura, 2003). Thus, the key components of the social cognitive theory are observational learning, reciprocal determinism, behavioral capacity, expectation, reinforcements, self and collective efficacy, and self-regulation. These components can individually or collectively lead to the adoption of particular practices.

With regard to social media skit usage, individuals exposed to skits learn new behaviors that they can potentially adopt in their regular lifestyles (Kim et al., 2020). The content from the skits can further get reinforced through social norms and cognitive processes (Kim et al., 2020; Kelly et al., 2017). The desire to conform and fit in one's social circle can compel individuals to practice what their peers have adopted through social media content sharing. The reinforcement of social media skit content, through behavioral user-engagement metrics such as likes, shares, and comments, can also drive the audiences to adopt the messages being promoted. This leads to the suggestion that

H3: Social media skits have some efficacy in influencing waste material separation behaviors

Materials and Methods

Research Philosophy and Design

A positivist epistemological research framework was adopted in seeking to achieve the objectives of this study. A positivist research philosophy seeks to establish causal relationships between variables (Irshaidat, 2022; Ruffa & Evangelista, 2021), usually through experimental research designs (Horváth, 2016). The conclusions reached in positivist research are based on empirical evidence through analyzing observable or measurable data in a quantitative format (Alharahsheh & Pius, 2020; Mohajan, 2020; Ayeni et al., 2019).

The research design adopted in this study was a quasi-experiment. A quasi-experimental research design seeks to establish whether the relationship between the study constructs follows a cause-and-effect pattern (Maciejewski, 2020; Reichardt, 2019; Kim & Steiner, 2016). Quasi-experiments are suitable for field settings due to challenges around ethical issues and random assignments needed for true experiments (Miller et al., 2020; Rogers & Revesz, 2019). Moreover, quasi-experiments are associated with higher external validity due to real-world interventions (Miller et al., 2020; Handley et al., 2018).

Population and Sampling

The population of this study was made up of baby caregiving mothers with children under the age of 24 months in the town of Bindura. The actual figure of baby caregiving mothers could not be ascertained due to limited availability of accurate official figures. However, on the basis of the need to achieve sufficient statistical power, a sample of 200 women was surveyed. Most of the mothers that become part of this study were married (55%) and were multiparous (60%). About 69% of the respondents were formally employed, while 22% were informally employed, with the remainder 9% being unemployed. The majority of the respondents were also formally educated (60%), while the remainder were either uneducated or informally educated (40%).

The target population was drawn from baby caring mothers who were contacted at the clinics in high-density surbarbs of the Bindura town. The intervention group was made up of those who were active on social media, while the control group was made up of those who were not active on social media. The respondents from both the groups were selected using convenience sampling. Ideally, a probability-based sampling method was the best, but some of the target respondents were not willing to participate in the study. Therefore, the researchers had to rely on those that were willing to take part in the study.

Measures

The measures used in this study were adapted from the extant literature. The adaptation of measurement scales from the extant literature was favored on the basis that it allows for the generation of results that could be compared to previous research. Waste separation at source awareness was assessed using three measures adapted from Wan et al. (2014), reading as follows: Waste separation reduces wasteful use of landfills (AWI), Waste separation conserves natural resources (AW2), Waste separation improves environmental quality (AW3). Waste separation at source attitudes were assessed using three measures adapted from Liao et al., 2018: Waste separation is rewarding (AT1), Waste separation is responsible (AT2), Waste separation is hygienic (AT3). Waste material separation at source practices were assessed using three measures adapted from Wan et al. (2014): I intend to separate my household waste in the near future

(IN1), I will separate my household waste every time I have it for disposal (IN2), and I am willing to participate in the separation scheme in the near future (IN3).

Data Collection Procedures

Data was collected from the women with babies of the ages below 24 months. Women were chosen as the respondents on the basis that according to the ecofeminism theory, they are the ones that suffer most from the brunt of poorly managed environments (Öztürk, 2020; Estévez-Saá & Lorenzo-Modia, 2018). The women selected were from the high densities of Bindura town. The high densities were chosen on the basis that they have a higher concentration of women with babies. Moreover, the high-density suburbs have got higher levels of indiscriminate disposal of used baby diapers as evidenced by the presence of several illegal dumpsites (Mangizvo et al., 2024; Muringaniza et al., 2023). The respondents were issued with a short questionnaire that assessed for exposure to the baby diaper disposal skit that was shared to the community through WhatsApp. The respondents were also requested to respond to the questions related to their levels of awareness, attitudes, and practices with regard to responsible disposal of used baby diapers, particularly in the form of waste material separation at source.

Data Analysis

Data analysis was conducted in two stages. The first stage related to assessing the measurement issues related to the latent variables as well as measurement invariance. Data analysis was conducted using a one-way multivariate analysis of variance (MANOVA). One-way MANOVA was suitable on the basis that the study had two groups, namely the experimental group and the control groups as well as three dependent variables. Prior to running a one-way MANOVA test, some related statistical assumptions were considered such as multivariate normality, *homogeneity of covariance matrices, and* equality of variance.

Results

The results are presented in two sections. The first section deals with the measurement issues and the second section deals with hypotheses testing.

Measurement Issues

The first measurement issue assessed was the model fit, by running a confirmatory factor analysis (CFA). The results indicated that the study has some acceptable level of fit as indicated by the Chisq/df of 3.406, which was above the threshold of 5 suggested by Kline (Kline, 2023). The other fit indices also supported the *model's ability to reproduce the* survey *data*, NFI = 0.944, RFI = 0.916, IFI = 0.960, TLI = 0.939, and CFI = 0.959. Having attained an acceptable model fit, the analysis proceeded to assessing construct validity and composite reliability.

Construct validity assessed was in the form of convergent and discriminant validity. The former was assessed using the Fornell and Larcker (1981) criterion based on average variance extracted (AVE) and the latter was determined by comparing each construct's AVE to pairs of shared variance (SV). Composite reliability (CR) was assessed using Hair et al. procedures. The results for construct validity and composite reliability are shown in Table 12.1.

The results shown in Table 12.1 indicated that there was sufficient convergent validity for all the constructs, as indicated by each construct loading significantly to its factor as well as the AVE for all the constructs being above the threshold of 0.5. There was also acceptable discriminant validity as indicated by the AVE for each construct being higher than the SV for different pairs. The CR for each construct was also acceptable, being higher than the threshold of 0.7.

The next analysis focused on measurement invariance. Measurement invariance assesses whether survey items have the same meaning across different groups (Cheung & Rensvold, 2002). In this study, measurement invariance was assessed at the configural level only using the multi-group confirmatory factor analysis (MG-CFA). The measurement invariance

Table 12.1 Reliability statistics

Construct	*CR*	*AVE*	*Shared Variance*		
			1	*2*	*3*
Waste separation awareness (1)	0.905	0.760	1		
Waste separation attitudes (2)	0.958	0.884	0.36	1	
Waste separation practice(practice (3)	0.840	0.637	0.07	0.07	1

Source: Primary Data

tests indicated that the three-factor model had an acceptable configural measurement invariance evidenced by adequate fit indices for the experimental groups, NFI = 0.902, IFI = 0.905, CFI = 0.905, and the control group, NFI = 0.922, IFI = 0.926, CFI = 0.926. The results revealed that the survey items are psychometrically equivalent across the experimental and the control groups through well-fitting baseline fit indices. This suggests that there was an identical conceptual framework for responding to the survey items (Specker, 2024).

Hypotheses Testing

One-way **MANOVA** *was conducted to determine whether social media influences various waste separation at source outcomes.* One-way MANOVA is a robust statistical tool that extends the capabilities of the analysis of variance (ANOVA) tool, thereby allowing for the simultaneous assessment of the impact of a single nominal independent variable on multiple metric dependent variables (McQuitty, 2018). *Prior to hypotheses testing, several MANOVA assumptions were assessed in order to ensure the statistical inferences attained were accurate and robust (*Field, 2024*).*

The first assumption assessed was multivariate normality. All the dependent variables combined should demonstrate multivariate normality. Multivariate normality was assessed using the Mahalnobis distance (D) statistic and the obtained results indicated sufficient multivariate normality since the critical value of 16.27 was not exceeded (max.value = 14.97).

The second assumption assessed was the homogeneity of covariance matrices, also known as the homogeneity of variance-covariance. The homogeneity of variance-covariance was assessed using Box's M test of equality of covariance *and the results indicated that* the covariance matrices were not equal, Box's $M = 41.492$, $F(6, 284{,}044.075) = 6.802$, $p < 0.001$. The violation of the equality of covariance assumption resulted in the adoption of the Pillai's Trace which is robust to this violation (Field, 2024).

The third assumption was the equality of variance and this was tested using Levine's test of equality of error variances. The variances were unequal for waste separation at source attitudes, $F(1, 198) = 2.466$, $p = 0.118$, and waste separation at source attitudes, $F(1, 198) = 4.285$, $p = 0.040$, except for waste separation at source attitudes, $F(1, 198) = 7.922$, $p = 0.005$. Since there were equal cases for all the constructs, homoscedasticity was deemed inconsequential. Having satisfied most of the one-way MANOVA requirements, the analysis proceeded to hypotheses testing and the results are shown in Table 12.2.

The default MANOVA hypotheses predicted that social media influences the multivariate pattern of waste separation at source for used baby diapers outcome variables. The results indicated that there are significant differences in the multivariate pattern of the combined dependent variables (waste separation at source awareness, attitude, and practice) across the levels of the social media skits. There was a statistically significant difference in waste separation outcomes based on social media skits, F (3, 196) = 54.226, $p < 0.001$; Wilk's $\Lambda = 0.546$, partial $\eta^2 = 0.454$.

Since the MANOVA test was significant, multiple ANOVA tests were run in order to test individual hypothesis. ANOVA tests function as post-hoc procedures after a significant MANOVA test result (Al-Abdullatif, 2020). An alpha correction was adjusted to (0.05/3) using the Bonferroni correction to cater for multiple ANOVAs being run. The results for multiple ANOVAs related to each hypothesis are shown in Table 12.3.

The first hypothesis suggested that there is a significant difference in the mean vectors of waste separation at source awareness across the levels of the social media skits. This hypothesis was supported, F (1, 198) = 53.380; $p < 0.001$; partial $\eta^2 = 0.212$, observed power = 1.000. The second hypothesis suggested that there is a significant difference in the mean vectors of waste separation at source attitude across the levels of the social media skits. This hypothesis was supported, $F(1, 198) = 56.662$; $p < 0.001$; partial $\eta^2 = 0.222$, observed power = 1.000. The third hypothesis suggested that there is a significant difference in the mean vectors of waste separation at source behavior across the levels of the social media skits. This hypothesis was supported, $F(1, 198) = 90.115$; $p < 0.001$; partial $\eta^2 = 0.313$, observed power = 1.000.

Discussion

The study sought to determine the effectiveness of social media skit in influencing awareness, attitudes, and practice of waste separation at source for used baby diapers. A social media skit was developed and shared through social media platforms. The results were compared between the experimental group and the control group. This is to the best knowledge of the researchers the first quasi-experimental study that involved the efficacy of social media skits as an intervention. Despite social media skits being sometimes criticized for encouraging sexual objectification of

Table 12.2 Multivariate Tests

Effect		*Value*	*F*	*Hypothesis df*	*Error df*	*Sig.*	*Partial Eta Squared*	*Observed Power*[c]
Intercept	Pillai's Trace	0.961	1630.761[b]	3.000	196.000	0.000	0.961	1.000
	Wilks' Lambda	0.039	1630.761[b]	3.000	196.000	0.000	0.961	1.000
	Hotelling's Trace	24.961	1630.761[b]	3.000	196.000	0.000	0.961	1.000
	Roy's Largest Root	24.961	1630.761[b]	3.000	196.000	0.000	0.961	1.000
Social Media	Pillai's Trace	0.454	54.226[b]	3.000	196.000	0.000	0.454	1.000
	Wilks' Lambda	0.546	54.226[b]	3.000	196.000	0.000	0.454	1.000
	Hotelling's Trace	0.830	54.226[b]	3.000	196.000	0.000	0.454	1.000
	Roy's Largest Root	0.830	54.226[b]	3.000	196.000	0.000	0.454	1.000

[a]Design: Intercept + Social Media

[b]Exact statistic

[c]Computed using alpha = 0.05

Table 12.3 Univariate tests

Source	*Dependent Variable*	*df*	*F*	*Sig.*	*Partial Eta Squared*	*Observed Power*[d]
Corrected Model	Waste separation at source awareness	1	53.380	0.000	0.212	1.000
	Waste separation at source attitudes	1	56.662	0.000	0.222	1.000
	Waste separation at source practice	1	90.115	0.000	0.313	1.000
Intercept	Waste separation at source awareness	1	2682.740	0.000	0.931	1.000
	Waste separation at source attitudes	1	2446.330	0.000	0.925	1.000
	Waste separation at source practice	1	1452.653	0.000	0.880	1.000
Social Media skits	Waste separation at source awareness	1	53.380	0.000	0.212	1.000
	Waste separation at source attitudes	1	56.662	0.000	0.222	1.000
	Waste separation at source practice	1	90.115	0.000	0.313	1.000
Error	Waste separation at source awareness	198				
	Waste separation at source attitudes	198				
	Waste separation at source practice	198				
Total	Waste separation at source awareness	200				
	Waste separation at source attitudes	200				
	Waste separation at source practice	200				
Corrected Total	Waste separation at source awareness	199				
	Waste separation at source attitudes	199				
	Waste separation at source practice	199				

[a] R Squared = 0.212 (Adjusted R Squared = 0.208)

[b] R Squared = 0.222 (Adjusted R Squared = 0.219)

[c] R Squared = 0.313 (Adjusted R Squared = 0.309)

[d] Computed using alpha = 0.05

women (Akalonu & Ha, 2024), their efficacy in promoting good social causes have been proved in this study and other previous studies in different social categories (e.g. Obiechina, 2023; Amonyeze et al., 2022; Ojomo & Sodeinde, 2021).

The results of this study indicated that social media skits significantly influence the levels of awareness for used baby diapers separation at source. More specifically, the results indicated that skits developed to raise environmental issues awareness are effective when shared through social media platforms such as WhatsApp. The use of social media skits shared through WhatsApp to promote environmental consciousness is a novel approach that hitherto has been limited to sharing basic social information only. Previous environmental consciousness information shared through social media has been limited to other social media platforms such as YouTube. Moreover, most information had been disseminated through documentaries that most people are likely to ignore due to their longer durations and highly formalized approach (Akben, 2019).

The results also revealed that social media skits are effective in inculcating positive attitudes towards sustainable used baby diaper disposal. These findings provide further empirical support to the effectiveness of social media usage, particularly through the development and dissemination of social media skits. It has already been indicated that social media skits have some influence on environmental attitudes (Omoyajowo et al., 2024), and this study added some empirical weight to that through focusing on used baby diapers.

Finally, the results indicated that social media skits are effective in inducing acceptable used baby diapers disposal practices. This suggests that when individuals are exposed to informational skits disseminated through social media platforms, they are likely to adopt and practice the promoted actions. This finding is in line with previous research where individuals adopted the practices that are promoted through social media skits (Zhao et al., 2023).

Theoretical Implications

The study has got some theoretical contributions particularly to the nudging theory through providing some empirical evidence in the form of robust data supporting the efficacy of nudging in behavioral changes. More specifically, the theoretical contribution of a study is in the provision of insights into how small changes in an intervention lead to significant alterations in behavior. The application of the nudging theory led to a proposal for a new intervention design that is easy to implement and cost-effective, such as modifying the extant environmental consciousness communication strategies. Thus, a fundamental theoretical contribution is providing empirical evidence supportive of the efficacy of nudges.

Managerial Implications

The findings from this study have significant implications for policymakers tasked with reducing or even eliminating indiscriminate disposal of domestic waste through practical interventions leveraging on behavioral insights. The demonstration of well-structured social media communications can efficiently nudge citizens towards a more environmentally responsible behavior. Finally, the findings in this research could inform future legislation as well as community programs aimed at enhancing environmental cleanliness.

Limitations of the Study

This study adopted a quasi-experimental design which has inherent limitations with regard to some inability to randomly assign participants to conditions. The absence of randomization is likely to account for the differences over and above the treatment introduced. This potentially introduces confounding effects and impacts negatively on internal validity issues. However, the study managed to attain some external validity through introducing social media skits which are real-world interventions.

Conclusions and Recommendations

The results from this study lead to a conclusion that social media skits lead to sustainable baby diapers disposable awareness, attitudes, and practices. These results also provide empirical validation to the overarching nudge theory where the provision of nudges brings about some noticeable changes in the intended outcomes. The predictive power of the nudge theory was demonstrated through support to all the hypothesized theories. Furthermore, theories that were used to develop individual hypothesis, such as the social amplification theory, cognitive social theory, and looking glass self-theory were empirically validated.

Practically, the results in this study suggest that environmentalists concerned with sustainable waste disposal must reach out to the general populace through social media. They must develop social media skits that resonate with the daily lives of the target audience. Therefore, health practitioners and environmental activists must direct their energy and resources to developing social media skits targeting baby caregiving mothers. This may involve the use of well-known socialites in order to promote user engagement.

References

Afolabi, B. E. I. F., & Falusi, T. A. (2023). Skits and comic illustrations: Means of transmedia storytelling and a platform for social change through healthy learning. *Multidisciplinary Journal of School Education, 12*(2 (24)), 297–317.

Agbefe, L. E., Lawson, E. T., & Yirenya-Tawiah, D. (2019). Awareness on waste segregation at source and willingness to pay for collection service in selected markets in Ga West Municipality, Accra, Ghana. *Journal of Material Cycles and Waste Management, 21*, 905–914.

Aichner, T., Grünfelder, M., Maurer, O., & Jegeni, D. (2021). Twenty-five years of social media: A review of social media applications and definitions from 1994 to 2019. *Cyberpsychology, Behavior, and Social Networking, 24*(4), 215–222.

Akalonu, C. P., & Ha, L. (2024). The role of online comedy skits in the sexual objectification of women in Nigeria. *Sexuality & Culture*, 1–17.

Akben, N. (2019). The views of students on the documentaries that they watched in history of science classes. *Social Sciences, 6*(5), 106–111.

Al-Abdullatif, F. (2020). *Discriminant function analysis versus univariate ANOVA as post hoc procedures following significant MANOVA test: A Monte Carlo study.* Ohio University.

Alharahsheh, H. H., & Pius, A. (2020). A review of key paradigms: Positivism VS interpretivism. *Global Academic Journal of Humanities and Social Sciences, 2*(3), 39–43.

Amonyeze, C., Okeke, A., & Agbo, O. (2022). Evaluation of Covid-19 themed comedy skits in social media health reportage in Southern Nigeria. *Webology, 19*(4), 34–48.

Atherton, J. (2023). *Social media strategy: A practical guide to social media marketing and customer engagement.* Kogan Page Publishers.

Ayeni, E. O., Saman, U. P., & Kasimu, S. (2019). Facts and fiction in positivism and neo positivism. *Research on Humanities and Social Sciences, 9*(4), 21–33.

Bammert, S., König, U. M., Roeglinger, M., & Wruck, T. (2020). Exploring potentials of digital nudging for business processes. *Business Process Management Journal, 26*(6), 1329–1347.

Bandura, A. (2001). Social cognitive theory: An agentic perspective. *Annual Review of Psychology, 52*(1), 1–26.
Bandura, A. (2003). Social cognitive theory for personal and social change by enabling media. In *Entertainment-education and social change* (pp. 97–118). Routledge.
Barker, H., Shaw, P. J., Richards, B., Clegg, Z., & Smith, D. (2021). What nudge techniques work for food waste behaviour change at the consumer level? A systematic review. *Sustainability, 13*(19), 11099.
Beauchamp, M. R., Crawford, K. L., & Jackson, B. (2019). Social cognitive theory and physical activity: Mechanisms of behavior change, critique, and legacy. *Psychology of Sport and Exercise, 42*, 110–117.
Braddock, K., & Morrison, J. F. (2021). Cultivating trust and perceptions of source credibility in online counternarratives intended to reduce support for terrorism. In *Narrative, Political Violence and Social Change* (pp. 41–66). Routledge.
Carr, C. T., & Hayes, R. A. (2015). Social media: Defining, developing, and divining. *Atlantic Journal of Communication, 23*(1), 46–65.
Chen, H. T., Kim, Y., & Chan, M. (2022). Just a glance, or more? Pathways from counter-attitudinal incidental exposure to attitude (de) polarization through response behaviors and cognitive elaboration. *Journal of Communication, 72*(1), 83–110.
Cheung, G. W., & Rensvold, R. B. (2002). Evaluating goodness-of-fit indexes for testing measurement invariance. *Structural Equation Modeling, 9*(2), 233–255.
Chong, M., & Choy, M. (2018). The social amplification of haze-related risks on the internet. *Health Communication, 33*(1), 14–21.
Cooley, C. H. (1902). *Human nature and the social order*. Scribner.
Coombs, W. T., & Holladay, S. J. (2021). Social amplification of scandals: One social media effect. *Scandology 3: Scandals in New Media*, 11–23.
Da Mota, V. T., & Pickering, C. (2020). Using social media to assess nature-based tourism: Current research and future trends. *Journal of Outdoor Recreation and Tourism, 30*, 100295.
Dangaiso, P., Jaravaza, D. C., & Mukucha, P. (2024). Evaluating the predictors of mobile health acceptance among Zimbabwean university students during the COVID-19 era: An integrated framework. *Cogent Social Sciences, 10*(1), 2299141.
Divya, B., Mehta, M., & Rayi, L. P. (2023, October). Use of instagram insights for increasing overall reach and engagement. In AIP conference proceedings (Vol. 2869, No. 1). AIP Publishing.
Estévez-Saá, M., & Lorenzo-Modia, M. J. (2018). The ethics and aesthetics of eco-caring: Contemporary debates on ecofeminism (s). *Women's Studies, 47*(2), 123–146.

Evans, D., Bratton, S., & McKee, J. (2021). *Social media marketing*. AG Printing & Publishing.

Faltesek, D., Graalum, E., Breving, B., Knudsen, E., Lucas, J., Young, S., & Varas Zambrano, F. E. (2023). TikTok as television. *Social Media+ Society, 9*(3), 20563051231194576.

Fellenor, J., Barnett, J., Potter, C., Urquhart, J., Mumford, J. D., & Quine, C. P. (2018). The social amplification of risk on Twitter: The case of ash dieback disease in the United Kingdom. *Journal of Risk Research, 21*(10), 1163–1183.

Field, A. (2024). *Discovering statistics using IBM SPSS statistics*. Sage publications limited.

Fornell, C., & Larcker, D. F. (1981). Evaluating structural equation models with unobservable variables and measurement error. *Journal of Marketing Research, 18*(1), 39–50.

Friedli, L. (2017). From the margins to the mainstream: The public health potential of mental health promotion–or mental health promotion–what works? In *Promotion of mental health* (pp. 77–88). Routledge.

Handley, M. A., Lyles, C. R., McCulloch, C., & Cattamanchi, A. (2018). Selecting and improving quasi-experimental designs in effectiveness and implementation research. *Annual Review of Public Health, 39*(1), 5–25.

Hasell, A. (2021). Shared emotion: The social amplification of partisan news on Twitter. *Digital Journalism, 9*(8), 1085–1102.

Horváth, I. (2016). Theory building in experimental design research. In *Experimental design research: Approaches, perspectives, applications* (pp. 209–231). Springer International Publishing.

Huang, X., Wang, S., Zhang, M., Hu, T., Hohl, A., She, B., & Li, Z. (2022). Social media mining under the COVID-19 context: Progress, challenges, and opportunities. *International Journal of Applied Earth Observation and Geoinformation, 113*, 102967.

Hudnut-Beumler, J., Po'e, E., & Barkin, S. (2016). The use of social media for health promotion in Hispanic populations: A scoping systematic review. *JMIR Public Health and Surveillance, 2*(2), e5579.

Hussain, F., Luby, S. P., Unicomb, L., Leontsini, E., Naushin, T., Buckland, A. J., & Winch, P. J. (2017). Assessment of the acceptability and feasibility of child potties for safe child feces disposal in rural Bangladesh. *The American Journal of Tropical Medicine and Hygiene, 97*(2), 469.

Irshaidat, R. (2022). Interpretivism vs. positivism in political marketing research. *Journal of Political Marketing, 21*(2), 126–160.

Itsubo, N., Wada, M., Imai, S., Myoga, A., Makino, N., & Shobatake, K. (2020). Life cycle assessment of the closed-loop recycling of used disposable diapers. *Resources, 9*(3), 34.

Iyanuoluwa, O., & Oyinlade, A. O. (2024). Switching the script: Unveiling the linguistic dance of code-switching in Nigerian online skits. *Linguistics Initiative, 4*(1), 155–173.

Jaravaza, D. C., Risiro, J., Mukucha, P., & Jaravaza, N. (2024). Public health promotion of COVID-19 vaccination to rural consumers: Synthesising the role of social media and religious belief systems. *International Journal of Pharmaceutical and Healthcare Marketing, 18*(1), 67–85.

John, P., Cotterill, S., Moseley, A., Richardson, L., Smith, G., Stoker, G., & Wales, C. (2019). *Nudge, nudge, think, think: Experimenting with ways to change citizen behaviour*. Manchester University Press.

Kasperson, R. E., Renn, O., Slovic, P., Brown, H. S., Emel, J., Goble, R., & Ratick, S. (1988). The social amplification of risk: A conceptual framework. *Risk Analysis, 8*(2), 177–187.

Kasperson, R. E., Webler, T., Ram, B., & Sutton, J. (2022). The social amplification of risk framework: New perspectives. *Risk Analysis, 42*(7), 1367–1380.

Kelly, M. E., Duff, H., Kelly, S., McHugh Power, J. E., Brennan, S., Lawlor, B. A., & Loughrey, D. G. (2017). The impact of social activities, social networks, social support and social relationships on the cognitive functioning of healthy older adults: A systematic review. *Systematic Reviews, 6*, 1–18.

Khanal, A., Giri, S., & Mainali, P. (2023). The practices of at-source segregation of household solid waste by the youths in Nepal. *Journal of Environmental and Public Health, 2023*(1), 5044295.

Khoo, S. C., Phang, X. Y., Ng, C. M., Lim, K. L., Lam, S. S., & Ma, N. L. (2019). Recent technologies for treatment and recycling of used disposable baby diapers. *Process Safety and Environmental Protection, 123*, 116–129.

Kim, H., Kim, Y., & Lee, D. (2020). Understanding the role of social media in political participation: Integrating political knowledge and bridging social capital from the social cognitive approach. *International Journal of Communication, 14*, 22.

Kim, Y., & Steiner, P. (2016). Quasi-experimental designs for causal inference. *Educational Psychologist, 51*(3–4), 395–405.

Kizgin, H., Dey, B. L., Dwivedi, Y. K., Hughes, L., Jamal, A., Jones, P., & Williams, M. D. (2020). The impact of social media on consumer acculturation: Current challenges, opportunities, and an agenda for research and practice. *International Journal of Information Management, 51*, 102026.

Kline, R. B. (2023). *Principles and practice of structural equation modeling*. Guilford Publications.

Kroll, T., & Stieglitz, S. (2021). Digital nudging and privacy: Improving decisions about self-disclosure in social networks. *Behaviour & Information Technology, 40*(1), 1–19.

Kuhl, J., Bisimwa, L., Thomas, E. D., Williams, C., Ntakirutimana, J., Coglianese, N., & George, C. M. (2021). Formative research for the development of baby water, sanitation, and hygiene interventions for young children in the Democratic Republic of the Congo (REDUCE program). *BMC Public Health, 21*, 1–16.

Langaro, D., Rita, P., & de Fátima Salgueiro, M. (2018). Do social networking sites contribute for building brands? Evaluating the impact of users' participation on brand awareness and brand attitude. *Journal of Marketing Communications, 24*(2), 146–168.

Le Dinh, C., Fujiwara, T., & Phu, S. T. P. (2021). Waste segregation at source and separately collection–the inadequacy in implementation in Vietnam. *Chemical Engineering Transactions, 89*, 499–504.

Li, R., Crowe, J., Leifer, D., Zou, L., & Schoof, J. (2019). Beyond big data: Social media challenges and opportunities for understanding social perception of energy. *Energy Research & Social Science, 56*, 101217.

Liao, C., Zhao, D., & Zhang, S. (2018). Psychological and conditional factors influencing staff's takeaway waste separation intention: An application of the extended theory of planned behavior. *Sustainable Cities and Society, 41*, 186–194.

Ling, C., Blackburn, J., De Cristofaro, E., & Stringhini, G. (2022, June). Slapping cats, bopping heads, and oreo shakes: Understanding indicators of virality in tiktok short videos. In Proceedings of the 14th ACM Web Science Conference 2022 (pp. 164–173).

Luthuli, S., Haskins, L., Mapumulo, S., Rollins, N., & Horwood, C. (2020). 'I decided to go back to work so I can afford to buy her formula': A longitudinal mixed-methods study to explore how women in informal work balance the competing demands of infant feeding and working to provide for their family. *BMC Public Health, 20*, 1–15.

Maciejewski, M. L. (2020). *Quasi-experimental design. Biostatistics & Epidemiology, 4*(1), 38–47.

Makoś-Chełstowska, P., Kurowska-Susdorf, A., & Płotka-Wasylka, J. (2021). Environmental problems and health risks with disposable baby diapers: Monitoring of toxic compounds by application of analytical techniques and need of education. *TrAC Trends in Analytical Chemistry, 143*, 116408.

Mangizvo, R. V., Lumbe, A., & Sibanda, A. (2024). The proposed ban of disposable baby diapers in Zimbabwe: Views from old senga residents, in the City of Gweru, Zimbabwe. *International Journal of Research in Education Humanities and Commerce, 5*(01), 38–53.

Martí, P., Serrano-Estrada, L., & Nolasco-Cirugeda, A. (2019). Social media data: Challenges, opportunities and limitations in urban studies. *Computers, Environment and Urban Systems, 74*, 161–174.

Mathe, M. (2018). Environmental pollution-perceptions and views on usage and disposa l of diapers: A case study of gwanda urban. *Environmental Pollution, 3*(5).

McQuitty, S. (2018). The purposes of multivariate data analysis methods: An applied commentary. *Journal of African Business, 19*(1), 124–142.

Mendoza, J. M. F., D'aponte, F., Gualtieri, D., & Azapagic, A. (2019). Disposable baby diapers: Life cycle costs, eco-efficiency and circular economy. *Journal of Cleaner Production, 211*, 455–467.

Miller, C. J., Smith, S. N., & Pugatch, M. (2020). Experimental and quasi-experimental designs in implementation research. *Psychiatry Research, 283*, 112452.

Mirbabaie, M., Ehnis, C., Stieglitz, S., Bunker, D., & Rose, T. (2021). Digital nudging in social media disaster communication. *Information Systems Frontiers, 23*, 1097–1113.

Mohajan, H. K. (2020). Quantitative research: A successful investigation in natural and social sciences. *Journal of Economic Development, Environment and People, 9*(4), 50–79.

Monge Roffarello, A., & De Russis, L. (2023, September). Nudging users towards conscious social media use. In *Proceedings of the 25th international conference on Mobile Human-Computer Interaction* (pp. 1–7).

Mukucha, P., Jaravaza, D. C., & Nyengerai, S. (2023a). Circular economy of shopping bags in emerging markets: A demographic comparative analysis of propensity to reuse plastic bags versus cotton bags and paper bags. *Cogent Engineering, 10*(1), 2176582.

Mukucha, P., Sixpence, S., Muzanenhamo, L., Muskwe, N., & Ukpere, W. I. (2023b). Green supply chain management: The role of sustainable packaging on corporate image and patronage intentions. *Annals of Spiru Haret University. Economic Series, 23*(4), 553–572.

Muringaniza, K., Madhara, K., Musasa, T., & Mupepi, O. (2023). Environmental challenges associated with disposal of baby diapers in Hwange town, Zimbabwe. *Journal of Policy and Society, 1*(1). https://doi.org/10.59400/jps.v1i1.284

Narayan, P., Jacob, A. M., Shetty, P., & Singh, P. (2022). A Comparative study on the knowledge, attitude, and perception toward hazardous medical waste produced at home in a Coastal City of Karnataka, India. *Journal of Health and Allied Sciences NU, 12*(02), 126–133.

Ng, Y. J., Yang, Z. J., & Vishwanath, A. (2018). To fear or not to fear? Applying the social amplification of risk framework on two environmental health risks in Singapore. *Journal of Risk Research, 21*(12), 1487–1501.

Ntekpe, M. E., Mbong, E. O., Edem, E. N., & Hussain, S. (2020). Disposable diapers: Impact of disposal methods on public health and the environment. *American Journal of Medicine and Public Health, 1*(2), 1009.

Nzima, D., & Maharaj, P. (2023). Motherhood, childcare and work: Exploring the multilayered challenges of finding paid work among young mothers in South Africa. *South African Review of Sociology, 53*(3), 313–327.

Obiechina, C. K. (2023). The mental health benefit of social media: The use of social media skits for depression treatment among women in Nigeria. *Ianna Journal of Interdisciplinary Studies, 5*(1), 1–10.

Ojomo, O., & Sodeinde, O. A. (2021). Social media skits: Reshaping the entertainment experience of broadcast audience. *Sage Open, 11*(3), 21582440211032176.

Okoye, C. M., & Okenyi, C. J. (2023). Rethinking the role of social media comedy on child development in Nigeria; a study of mark angel skits. *Ama: Journal of Theatre and Cultural Studies, 16*(1).

Omoyajowo, K. O., Raimi, M. O., Omoyajowo, K. A., Makengo, M. B., Adegboyo, S., Innocent, D. C., & Kakwi, D. (2024). Towards a reduced pollution society: Systematic review on the role of storytelling, social media, humor and celebrities' influence for research communication. *Journal of Applied Sciences and Environmental Management, 28*(2), 603–623.

Ortiz, R. R., Smith, A., & Coyne-Beasley, T. (2019). A systematic literature review to examine the potential for social media to impact HPV vaccine uptake and awareness, knowledge, and attitudes about HPV and HPV vaccination. *Human Vaccines & Immunotherapeutics, 15*(7–8), 1465–1475.

Öztürk, Y. M. (2020). An overview of ecofeminism: Women, nature and hierarchies. *Journal of Academic Social Science Studies, 13*(81), 705–714.

Płotka-Wasylka, J., Makoś-Chełstowska, P., Kurowska-Susdorf, A., Treviño, M. J. S., Guzmán, S. Z., Mostafa, H., & Cordella, M. (2022). End-of-life management of single-use baby diapers: Analysis of technical, health and environment aspects. *Science of the Total Environment, 836*, 155339.

Post, S. (2019). Polarizing communication as media effects on antagonists. Understanding communication in conflicts in digital media societies. *Communication Theory, 29*(2), 213–235.

Purwati, S., Sihono Gabriel, D., & Dachyar, M. (2021, May). Analysis of willingness to accept and factors affecting compensation for disposable baby diaper waste sorting in West Jakarta. In *Proceedings of the 4th Asia Pacific Conference on Research in Industrial and Systems Engineering* (pp. 448–452).

Randles, J. (2021). "Willing to do anything for my kids": Inventive mothering, diapers, and the inequalities of carework. *American Sociological Review, 86*(1), 35–59.

Reese, H., Alman, B., & Null, C. (2015). Disposing of children's diapers with solid waste: A global concern? *Waterlines, 34*(3), 255–268.

Reichardt, C. S. (2019). *Quasi-experimentation: A guide to design and analysis.* Guilford Publications.

Ren, S., & Liu, D. (2024). Effects of digital nudging on the adoption of APPs for value co-creation among online consumers. *Information Technology and Management, 25*(2), 161–172.

Rogers, J., & Revesz, A. (2019). Experimental and quasi-experimental designs. In *The Routledge handbook of research methods in applied linguistics* (pp. 133–143). Routledge.

Ruffa, C., & Evangelista, M. (2021). Searching for a middle ground? A spectrum of views of causality in qualitative research. *Italian political science review/Rivista italiana di scienza politica, 51*(2), 164–181.

Ruiz-Junco, N., & Brossard, B. (2019). *Updating Charles H. Cooley.* Routledge.

Sachidhanandham, A., & Priyanka, M. (2020). A review on convenience and pollution caused by baby diapers. *Science and Technology Development Journal, 23*(3), 699–712.

Schenck, C. J., Chitaka, T. Y., Tyrrell, H., & Couvert, A. (2023). Disposable diaper usage and disposal practices in Samora Machel Township, South Africa. *Sustainability, 15*(12), 9478.

Schivinski, B., Langaro, D., & Shaw, C. (2019). The influence of social media communication on consumer's attitudes and behavioral intentions concerning brand-sponsored events. *Event Management, 23*(6), 835–853.

Schunk, D. H., & DiBenedetto, M. K. (2020). Motivation and social cognitive theory. *Contemporary Educational Psychology, 60*, 101832.

Sheak, E., & Abdulrazak, S. (2023). The influence of social media marketing activities on TikTok in raising brand awareness. *Market-Tržište, 35*(1), 93–110.

Song, H., & Boomgaarden, H. G. (2017). Dynamic spirals put to test: An agent-based model of reinforcing spirals between selective exposure, interpersonal networks, and attitude polarization. *Journal of Communication, 67*(2), 256–281.

Specker, E. (2024). Further validating the VAIAK: Defining a psychometric model, configural measurement invariance, reliability, and practical guidelines. *Psychology of Aesthetics, Creativity, and the Arts, 18*(3), 449.

Thaler, R. H., & Sunstein, C. R. (2021). *Nudge: The final edition.* Yale University Press.

Tuten, T. L. (2023). *Social media marketing.* Sage publications limited.

Vafeiadis, M., Bortree, D. S., Buckley, C., Diddi, P., & Xiao, A. (2020). Refuting fake news on social media: Nonprofits, crisis response strategies and issue involvement. *Journal of Product & Brand Management, 29*(2), 209–222.

Wambui, K. E., Joseph, M., & Makindi, S. (2015). Soiled diapers disposal practices among caregivers in poor and middle income urban settings. *International Journal of Scientific and Research Publications, 5*(10), 154–163.

Wan, C., Shen, G. Q., & Yu, A. (2014). The role of perceived effectiveness of policy measures in predicting recycling behaviour in Hong Kong. *Resources, Conservation and Recycling, 83*, 141–151.

Waterhouse, P., Hill, A. G., & Hinde, A. (2017). Combining work and child care: The experiences of mothers in Accra. *Ghana. Development Southern Africa, 34*(6), 771–786.

White, H. L., Mwapasa, T., Mphasa, M., Kalonde, P. K., Feasey, N., Oliver, D. M., & Quilliam, R. S. (2023). Open defaecation by proxy: Tackling the increase of disposable diapers in waste piles in informal settlements. *International Journal of Hygiene and Environmental Health, 250*, 114171.

Wirz, C. D., Xenos, M. A., Brossard, D., Scheufele, D., Chung, J. H., & Massarani, L. (2018). Rethinking social amplification of risk: Social media and Zika in three languages. *Risk Analysis, 38*(12), 2599–2624.

Zhang, X. A., & Cozma, R. (2022). Risk sharing on Twitter: Social amplification and attenuation of risk in the early stages of the COVID-19 pandemic. *Computers in Human Behavior, 126*, 106983.

Zhao, H., Xu, J., Iyendo, T. O., Apuke, O. D., Tunca, E. A., & Gever, V. C. (2023). The effectiveness of using audio-visual based media intervention for promoting social media literacy skills to curtail fake news on social media: A quasi-experimental investigation. *Information Development*, 02666669231217236.

Zhao, S. (2018). Bringing self-values back. In *From reflected appraisal to appraised appraisal. In Updating Charles H. Cooley* (pp. 126–140). Routledge.

Zhao, S., & Wu, X. (2021). From information exposure to protective behaviors: Investigating the underlying mechanism in COVID-19 outbreak using social amplification theory and extended parallel process model. *Frontiers in Psychology, 12*, 631116.

CHAPTER 13

Sustainable Green Marketing Strategies for a Circular Economy in Africa: A Conclusion

Brighton Nyagadza and Asphat Muposhi

Overview of Green Marketing Realities in Africa

The status of green marketing implementation and achievements in Africa point to a bleak future. There is general consensus in extant literature that the African continent is lagging behind in attaining environmental sustainability targets (Asongu & Odhiambo, 2021; Aluko et al., 2023). Empirical evidence also suggests that the African continent is warming more rapidly

B. Nyagadza
School of Business and Health Studies, York St. John University (YSJU), London, UK
e-mail: b.nyagadza@yorksj.ac.uk

A. Muposhi (✉)
Marketing Division, School of Business Sciences, University of the Witwatersrand, Johannesburg, South Africa
e-mail: asphat.muposhi@wits.ac.za

A. Muposhi, B. Nyagadza (eds.), *Sustainable Green Marketing Strategies for a Circular Economy*, Palgrave Studies of Marketing in Emerging Economies,
https://doi.org/10.1007/978-3-031-81627-7_13

as compared to other continents (Avom et al., 2020; Aluko et al., 2023). This, according to Avom et al. (2020), is manifested by the escalation of climate change-induced grand problems such as droughts, floods, poverty, forced migration, health pandemics and social inequalities. Moreover, the growing rate of unemployment characterised by de-industrialisation and rapid informalisation of African economies is at odds with the green marketing promise of green economic growth (Asongu & Odhiambo, 2021). This further raises doubt about the potential of green marketing as a driver of economy growth in Africa.

At the macro level, the implementation of green marketing is constrained by lack of commitment from national governments (Asongu & Odhiambo, 2021; Sabela-Rikhotso et al., 2021). In the majority of African countries, a transition to a green economy is regarded as not urgent and is perceived as an antithesis to economic growth (Sabela-Rikhotso et al., 2021). Politics of sustainability dominate the international debates on green marketing between the Global North and Global South (Turkcu & Tura, 2023). For example, the setting of net-zero targets and the associated strategies remains a subject of contestation (Musah et al., 2024). Often, the commitment of national governments to green marketing issues depends on the political ideology of the governing party (Asongu & Odhiambo, 2021). Lack of green marketing policies and environmental standards as well as lapses in monitoring and enforcement of environmental laws are identified as major impediments (Asongu & Odhiambo, 2021). Poor levels of political governance characterised by corruption, lack of rule of law, conflict and human rights abuse are identified as grand challenges that inhibit the promotion of environmental citizenship behaviours in Africa (Aluko et al., 2023).

Emerging Imperatives for Green Marketing Implementation in Africa

Ubuntu Philosophy as Green Marketing Framework

This book provides valuable insights on green marketing governance in Africa. Due to the limited availability of resources required to enforce and monitor green marketing regulations, the adoption of the African philosophy of Ubuntu is recommended as a framework for promoting self-regulated pro-environmental behaviour. The Ubuntu philosophy has the

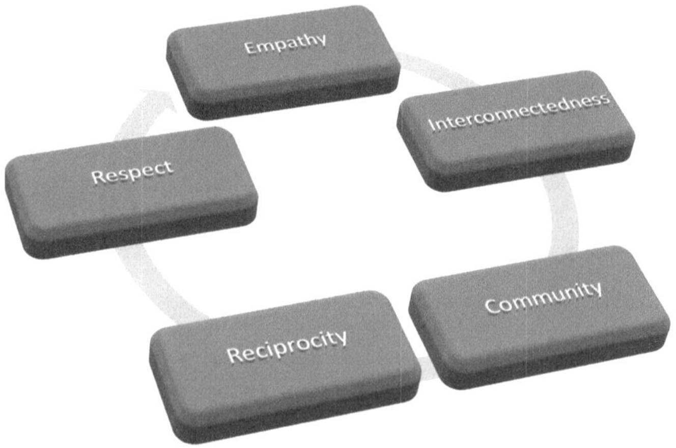

Fig. 13.1 Ubuntu green marketing framework

potential to contribute to good green marketing governance because it is underpinned by shared African values, beliefs and norms (Kelbessa, 2015; Terblanché-Greeff, 2019). Moreover, the Ubuntu philosophy is regarded as a more African-centric approach that decolonise the African environmental governance from dominant Western environmental management perspectives (Terblanché-Greeff, 2019; Maziriri et al., 2023). This African-centric approach is expected to result in transformative green marketing initiatives that yield mutual benefits for businesses and communities. The proposed green marketing framework is premised on the principles of Ubuntu such as respect, empathy, interconnectedness, reciprocity and community as follows (Fig. 13.1).

Green Marketing as a Business Model

The potential of green marketing as a business model was confirmed, including its ability to enable companies to reduce their environmental footprint while enhancing brand reputation and driving long-term growth. The case studies analysed showed that commitment to a green marketing strategy has resulted in improved transparency in sustainability reporting, corporate image, cost savings, stakeholder engagement and responsible investment. However, there is a need to adopt a systems approach,

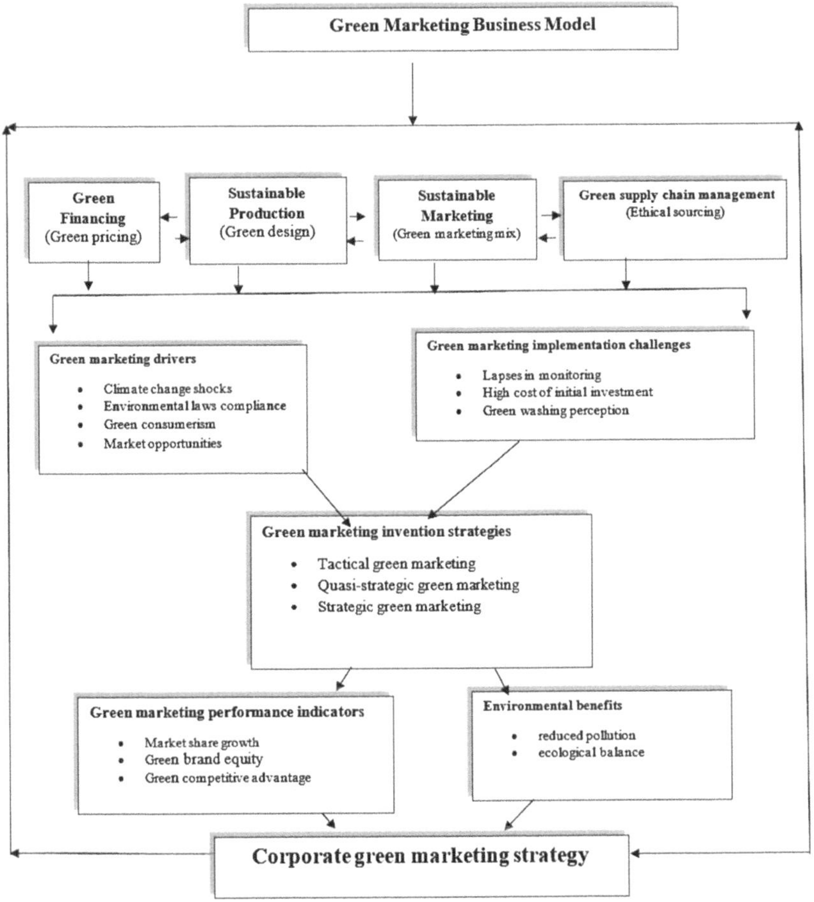

Fig. 13.2 Green marketing business model

evaluate the effectiveness of green marketing strategies and adopt green marketing standards to allay greenwashing. This book contributes to green marketing literature by proposing a green marketing business model (Fig. 13.2).

Managing Green Capitalism Perceptions

It emerged that consumers perceive that green marketing is evolving into 'green capitalism'. As a result of this perception, consumers exhibit ambivalent behaviour towards green marketing practices. Multinational companies take advantage of Africa's dire need for investment and lapses in enforcement of environmental regulations (Naidoo & Gasparatos, 2023; Nyagadza, 2021). As a result, in the majority of African countries, green marketing is mainly implemented under the template of self-regulation. Thus, there is a growing perception that companies are selectively implementing green marketing initiatives that are more aligned towards profitability than environmental sustainability. The majority of organic products are premium priced and beyond the reach of low-income groups, and this is regarded as a new form of market exclusion. Retailers in the grocery sector are also accused of making profit from the sale of reusable shopping bags. A lifecycle assessment of some of the reusable shopping bags showed that they are not environmentally friendly as claimed, thereby raising greenwashing concerns. Thus, the introduction of green marketing standards and consistent monitoring of the implementation of green marketing practices are key imperatives.

Bridging the Intention-Behaviour Gap

The prevalence of the intention-behaviour gap characterises consumers' behavioural response towards green products and social eco-innovations (Mkhize & Ellis, 2020; Udeagha & Muchapondwa, 2023). The gap is reflected by consumers who report favourable intentions to engage in pro-environmental behaviour yet fail to perform actual behaviour (Mukonza & Swarts, 2020). Mkhize & Ellis, 2024). To promote the mainstream consumption of green products and eco-innovations, there is a need to facilitate affordability through nudges and subsidies. Policymakers may also leverage on the power of artificial intelligence and social media to disseminate information on green products and eco-innovations.

Conclusion

This book explored African realities and imperatives associated with the implementation of green marketing strategies. The African philosophy of Ubuntu is recommended as a framework for promoting self-regulated green marketing practices. By unlocking the power of Ubuntu philosophy and indigenous wisdom, policymakers can develop more effective green marketing strategies that promote environmental justice and improve quality of life. The case studies analysed showed that commitment to a green marketing strategy is associated with improved transparency in sustainability reporting, corporate image, cost savings, stakeholder engagement and responsible investment that considers environmental, social and governance factors. However, there is a need to adopt a systems approach in the implementation of green marketing strategies in order to tap into the associated synergies. The role of sustainability indicators as tools for evaluating green marketing strategies was affirmed. Through the use of sustainability indicators, firms can make well-informed decisions, improve environmental performance and gain consumer trust. It emerged that consumers perceive that green marketing is evolving into 'green capitalism'. As a result of this perception, consumers are exhibiting ambivalent behaviour towards green marketing practices. Policymakers should consider implementing green marketing standards and consistently monitor the operations of companies to minimise greenwashing concerns.

References

Aluko, O. O., Opoku, E. O., & Acheampong, A. O. (2023). Economic complexity and environmental degradation: Evidence from OECD countries. *Business Strategy and the Environment, 32*(6), 2649–4014.

Asongu, S. A., & Odhiambo, N. M. (2021). Enhancing governance for environmental sustainability in sub-Saharan Africa. *Energy Exploration & Exploitation, 39*(1), 444–463.

Avom, D., Nkengfack, H., Fotio, H. K., & Totouom, A. (2020). ICT and environmental quality in Sub-Saharan Africa: effects and transmission channels. *Technological Forecasting and Social Change,* 155 (120028).

Kelbessa, W. (2015). Climate ethics and policy in Africa. *Thought and Practice: A Journal of the Philosophical Association of Kenya, 7*, 41–84.

Maziriri, E. T., Nyagadza, B., Chuchu, T., & Mazuruse, G. (2023). Antecedents that influence attitudes towards the use of environmentally friendly household appliance products in Zimbabwe, *PSU Research Review (PRR)*, Vol. ahead-of-print No. ahead-of-print. https://www.emerald.com/insight/content/doi/10.1108/PRR-03-2022-0033

Mkhize, S., & Ellis, D. (2020). Creativity in marketing communication to overcome barriers to organic produce purchases: The case of a developing nation. *Journal of Cleaner Production*, 242, (118415).

Mkhize, S., & Ellis, D. (2024). Organic consumption as a means to achieve sustainable development goals and agenda 2063. *Sustainable Development*, 1–12.

Mukonza, C., & Swarts, I. (2020). The influence of green marketing strategies on business performance and corporate image in the retail sector. *Business Strategy and the Environment, 29*(3), 838–845.

Musah, M., Onifade, S. T., Ankrah, I., Gyamfi, B. A., & Amoako, G. K. (2024). Achieving net-zero emission target in Africa: Are sustainable energy innovations and financialization crucial for environmental sustainability of sub-Saharan African state? *Applied Energy, 364*(123120).

Naidoo. M., & Gasparatos, A. (2023). Consumer worldviews and perspectives on environmental sustainability initiatives in the South African supermarket sector. *Journal of Cleaner Production*, 413 (137496).

Nyagadza, B. (2021). Fostering green economies in Africa through green marketing strategies for environmental sustainability: an overview. *Journal of Environmental Media (JEM), 2*(1), 17–22.

Sabela-Rikhotso, P. T. Z., van Niekerk, D., & Nemakonde, L. D. (2021). A critical analysis of the legal frameworks governing oil spill management in South Africa. *Marine Policy,* 127(104433).

Terblanché-Greeff, A. C. (2019). Ubuntu and environmental ethics: The West can learn from Africa when faced with climate change. In M. Chemhuru (Ed.), *African environmental ethics. The international library of environmental, agricultural and food ethics* (Vol. 29). Springer.

Turkcu, D., & Tura, N. (2023). The dark side of sustainable packaging: Battling with sustainability tensions. *Sustainable Production and Consumption, 40*, 412–421.

Udeagha, M. C., & Muchapondwa, E. (2023). Environmental sustainability in South Africa: Understanding the criticality of economic policy uncertainty, fiscal decentralization, and green innovation. *Sustainable Development, 31*(3), 16381651.

Index

A. Muposhi, B. Nyagadza (eds.), *Sustainable Green Marketing Strategies for a Circular Economy*, Palgrave Studies of Marketing in Emerging Economies,
https://doi.org/10.1007/978-3-031-81627-7

The manufacturer's authorised representative in the EU is Springer Nature Customer Service Centre GmbH, Europaplatz 3, 69115 Heidelberg, Germany. If you have any concerns regarding our products, please contact ProductSafety@springernature.com

Printed and bound by CPI Group (UK) Ltd, Croydon, CR0 4YY
01/10/2026
02225912-0008